UNDERSTANDING THE MECHANISMS OF CRISPR IMMUNITY IN PROKARYOTES

RICKBED NANDI

Preface

In the realm of molecular biology, a groundbreaking revolution has unfolded over the past few decades, forever altering our understanding of genetic manipulation and microbial defence mechanisms. At the heart of this scientific transformation lies a phenomenon known as CRISPR-Cas immunity in prokaryotes. This book, *"Understanding the Mechanisms of CRISPR Immunity in Prokaryotes,"* is an exploration of this remarkable and intricate biological defence system.

The journey into the depths of CRISPR-Cas immunity is both a tale of scientific discovery and a testament to the awe-inspiring complexity of life at its most fundamental level. It is a story that transcends laboratory benches and delves into the genomes of the tiniest life forms on our planet, prokaryotes. These microorganisms, which include bacteria and archaea, have developed an astonishing array of strategies to fend off viral invaders and maintain the integrity of their genetic code.

As we delve into the pages of this book, we will embark on a comprehensive voyage through the universe of CRISPR-Cas immunity. We will begin with the basics, providing an introduction to the concept and its evolutionary significance. From there, we will journey into the heart of the CRISPR-Cas system, dissecting its various components and elucidating their roles in the immune defence of prokaryotes.

The book will navigate through the intricate processes of CRISPR adaptation and interference, explaining how prokaryotes "remember" past encounters with invaders and subsequently target and destroy them. We will unravel the diversity of CRISPR-Cas systems, exploring the unique strategies adopted by different prokaryotic lineages. Moreover, we will unveil the ever-evolving arms race between prokaryotes and their viral adversaries, shedding light on the mechanisms of CRISPR evasion and the countermeasures deployed by viruses.

As we delve deeper, we will uncover the structural insights that have illuminated the inner workings of CRISPR-Cas systems and how these structures have shaped our understanding of their mechanisms. We will also examine the broader implications of CRISPR-Cas systems, from their applications in biotechnology and medicine to their impact on microbial ecology and the environment.

Throughout this journey, we will confront the challenges and ethical considerations inherent in CRISPR-Cas research, offering a glimpse into the complexities and controversies that surround this revolutionary field.

This book is not just a compilation of facts; it is a testament to the tireless efforts of countless scientists who have dedicated their careers to unravelling the mysteries of CRISPR-Cas immunity. It is an invitation to join the ranks of those who seek to understand and harness the power of this biological phenomenon.

Whether you are a seasoned researcher, a curious student, or a concerned citizen interested in the intersection of science and society, this book aims to provide you with a comprehensive and accessible resource. It is our hope that it will inspire you to explore, question, and contribute to the ever-evolving story of CRISPR-Cas immunity in prokaryotes.

So, let us embark on this journey together, as we uncover the mechanisms that underlie one of the most extraordinary natural defence systems that life on Earth has ever devised. Welcome to *"Understanding the Mechanisms of CRISPR Immunity in Prokaryotes."*

Rickbed Nandi

Contents

Chapter 1: Introduction to CRISPR Immunity

1.1 The Evolutionary Significance of CRISPR-Cas Systems

The evolutionary significance of CRISPR-Cas systems is a topic of great interest and has profound implications for our understanding of the coevolutionary dynamics between prokaryotes and their viral adversaries. To appreciate this significance fully, we must delve into the deep history of these remarkable immune systems.

Origins of CRISPR-Cas Systems

CRISPR (Clustered Regularly Interspaced Short Palindromic Repeats) and Cas (CRISPR-associated) systems have origins dating back to the early days of life on Earth. While the modern understanding of CRISPR-Cas systems has been largely shaped by the molecular biology revolution of the late 20th century, these systems have been evolving and adapting for billions of years.

One of the most intriguing aspects of CRISPR-Cas systems is their presence in the genomes of many different prokaryotic species, from bacteria to archaea. This wide distribution suggests that these systems are ancient and have been conserved throughout evolution.

Fossil records of ancient microorganisms, some dating back over 3 billion years, have revealed structures resembling modern stromatolites. These structures contain layers of microbial mats, suggesting that life existed in complex communities at a time when the Earth's atmosphere was drastically different from today. It's plausible that these ancient microorganisms had their own primitive defence mechanisms against viral threats, which could be early precursors to CRISPR-like systems.

Coevolutionary Arms Race

The evolutionary significance of CRISPR-Cas systems becomes especially apparent when considering the ongoing coevolutionary arms race between prokaryotes and their viral adversaries, such as bacteriophages (phages). Phages are viruses that specifically infect bacteria, and they are the most abundant biological entities on the planet.

Prokaryotes and phages have been locked in this arms race for eons. When a phage infects a bacterium, it injects its genetic material into the host cell, hijacking the cell's machinery to replicate itself. Prokaryotes, in response, have developed a variety of mechanisms to defend against phage infections, with CRISPR-Cas systems being one of the most sophisticated.

Studies have shown that phages can rapidly evolve to escape recognition by CRISPR-Cas systems. Researchers have documented instances where phages mutate their DNA sequences to avoid being targeted by the CRISPR-Cas immune system. In turn, prokaryotes adapt by acquiring new spacers that match the evolved phage sequences. This continuous cycle of adaptation and counter-adaptation underscores the intense selective pressure at play in the prokaryote-phage interaction.

Diversity of CRISPR-Cas Systems

Another piece of evidence highlighting the evolutionary significance of CRISPR-Cas systems is the incredible diversity of these systems. There are two main classes of CRISPR-Cas systems, Class 1 and Class 2, each with multiple subtypes, and a growing number of variations continue to be discovered.

Class 1 CRISPR-Cas systems are characterized by their multisubunit effector complexes and diverse set of Cas proteins. In contrast, Class 2 systems, which include the well-known Cas9 system, have a simpler single-protein effector. The existence of these distinct classes and subtypes suggests that CRISPR-Cas systems have undergone extensive diversification and specialization to meet the unique challenges posed by different phages and other invading genetic elements.

CRISPR as an Adaptive Immune System

One of the most intriguing aspects of CRISPR-Cas systems is their capacity for adaptive immunity. Unlike many other defence mechanisms in prokaryotes, which rely on pre-existing defences or general recognition mechanisms, CRISPR-Cas systems can learn from past encounters with invaders and tailor their responses accordingly.

Consider a scenario where a bacterium is infected by a new phage. Initially, the bacterium lacks specific immunity to this phage. However, during the infection, the bacterium incorporates a small piece of the phage's genetic material into its CRISPR array as a new spacer. This spacer serves as a memory of the phage, allowing the bacterium to mount a more effective defence if the same phage attacks in the future. This adaptive capability is akin to the way our immune system generates antibodies against novel pathogens.

Implications for Evolutionary Biology

The evolutionary significance of CRISPR-Cas systems extends beyond the realm of prokaryotes. Understanding the coevolutionary dynamics between prokaryotes and phages has broader implications for evolutionary biology.

CRISPR-Cas systems are not confined to prokaryotes. In some instances, genes related to CRISPR-Cas systems have been found in the genomes of eukaryotic organisms, suggesting potential horizontal gene transfer events between prokaryotes and eukaryotes. This raises questions about the role of CRISPR-Cas systems in shaping the evolution of more complex life forms.

Thus, the evolutionary significance of CRISPR-Cas systems is multifaceted. These systems have ancient origins, are deeply entwined in the ongoing coevolutionary struggle between prokaryotes and phages, showcase remarkable diversity, and provide insights into the adaptive capabilities of microorganisms. Beyond their immediate applications in biotechnology and medicine, CRISPR-Cas systems offer a window into the intricate dance of life on our planet and its ongoing evolutionary processes.

1.2 Prokaryotic Immune Systems: A Historical Overview

Prokaryotes, the simplest and most ancient life forms on Earth, have had to develop robust defence mechanisms to survive in a world filled with viruses and other potentially lethal invaders. This subsection provides a historical overview of prokaryotic immune systems, highlighting their evolution and the pivotal role they play in the survival and adaptation of these microorganisms.

The Dawn of Prokaryotic Immune Systems

The concept of prokaryotic immune systems dates back billions of years, long before the term "immune system" even existed. At their core, these systems are designed to protect prokaryotic cells from viral infections and other foreign genetic material. One of the earliest documented examples of prokaryotic immunity is the restriction-modification system.

The Restriction-Modification System

In the 1950s, scientists discovered that certain strains of Escherichia coli (E. coli) were resistant to infection by specific phages. This resistance was attributed to a mechanism now known as the restriction-modification system. It was found that E. coli cells could recognize and cleave foreign DNA using enzymes called restriction endonucleases, preventing the phage from replicating within the host cell. Simultaneously, the host DNA was chemically modified to protect it from the same enzymes. This system served as a precedent for the idea that prokaryotes possess innate defences against foreign DNA.

CRISPR-Cas: The Modern Era of Prokaryotic Immunity

The most significant leap in our understanding of prokaryotic immunity occurred with the discovery of Clustered Regularly Interspaced Short Palindromic Repeats (CRISPR) and CRISPR-associated (Cas) proteins. This breakthrough, which began in the late 1980s but gained significant momentum in the 2000s, revealed a sophisticated and adaptable immune system within prokaryotes.

The First Glimpses of CRISPR

The term CRISPR was first coined in a 2002 publication by Jansen et al., but the function of these mysterious repeating sequences remained unknown. It wasn't until 2005 that the groundbreaking work of Barrangou and colleagues demonstrated that CRISPR regions act as an adaptive immune system in Streptococcus thermophilus, a bacterium used in yogurt production. They showed that S. thermophilus could acquire short DNA sequences from phages that had previously attacked it and incorporate them into its own genome within the CRISPR loci.

Unravelling CRISPR Mechanisms

The breakthroughs continued in 2010, when Jennifer Doudna and Emmanuelle Charpentier, among others, elucidated the role of Cas9, a protein associated with CRISPR, in the cleavage of target DNA. They developed the CRISPR-Cas9 system as a revolutionary genome editing tool, allowing precise modifications of DNA in various organisms, including humans. This discovery earned them the Nobel Prize in Chemistry in 2020.

The Historical Significance of CRISPR-Cas

The discovery of CRISPR-Cas systems has fundamentally changed the way we view prokaryotic immunity and our ability to manipulate genetic material. It has evolved from a rudimentary restriction-modification system to a precise and adaptable immune system that prokaryotes use to fend off invasive genetic elements.

Beyond CRISPR: Other Prokaryotic Defence Mechanisms

While CRISPR-Cas systems have captured the spotlight in recent years, it's important to note that prokaryotes employ a variety of other defence mechanisms. These mechanisms, often working in conjunction with CRISPR, provide additional layers of protection against invaders.

Toxin-Antitoxin Systems

Toxin-antitoxin systems are genetic modules present in many bacterial genomes. They consist of a toxic protein and its cognate antitoxin. Under normal conditions, the antitoxin neutralizes the toxin. However, during stress or environmental changes, such as phage attack, the antitoxin can be degraded, allowing the toxin to harm the cell, potentially leading to growth arrest or death. Toxin-antitoxin systems are thought to help bacteria survive in adverse conditions and may play a role in defence against phage infections.

Restriction-Modification Systems

As mentioned earlier, restriction-modification systems were among the first recognized prokaryotic defence mechanisms. These systems typically involve restriction endonucleases that can cleave foreign DNA and modification enzymes that methylate the host DNA to protect it from cleavage. While not as adaptable as CRISPR-Cas systems, restriction-modification systems remain important contributors to prokaryotic immunity.

Prophage Activation and Superinfection Exclusion

Prokaryotes can also defend against phages by activating prophages, which are dormant phage genomes integrated into the bacterial chromosome. Activating a prophage can lead to production of phage particles that target and kill incoming phages. Additionally, some prokaryotes employ a mechanism known as superinfection exclusion, where they prevent secondary infections by closely related phages after the first infection. These strategies help prokaryotes resist repeated phage attacks.

Prokaryotic immune systems have come a long way from their humble beginnings as basic restriction-modification systems. The discovery of CRISPR-Cas systems, in particular, has revolutionized our understanding of how prokaryotes defend themselves against invaders. These systems, with their ability to adapt to new threats, are not only fascinating examples of biological innovation but also invaluable tools for biotechnology and medicine. As we delve deeper into the mechanisms of CRISPR-Cas immunity in prokaryotes in the subsequent chapters, it's important to keep in mind the historical context in which these remarkable systems were discovered and their significant impact on science and technology.

1.3 Scope and Objectives of the Book

The scope and objectives of this book on "Understanding the Mechanisms of CRISPR Immunity in Prokaryotes" are multifaceted, aiming to provide a comprehensive understanding of the intricate world of prokaryotic immunity through CRISPR-Cas systems. In this section, we delve into the specific goals and boundaries of this book, outlining what readers can expect to learn from its chapters.

Defining the Scope

Prokaryotes, which include bacteria and archaea, are among the most abundant and diverse life forms on Earth. Despite their small size and simple structure, they possess sophisticated defence mechanisms that allow them to fend off a myriad of threats, particularly from viruses called bacteriophages. One of the most remarkable of these defences is the Clustered Regularly Interspaced Short Palindromic Repeats (CRISPR) system, accompanied by CRISPR-associated (Cas) proteins. This book centres its scope primarily on the mechanisms, evolution, and applications of CRISPR immunity in prokaryotes.

Objectives of the Book

The main objectives of this book are as follows:

Objective 1: Unveiling the Complexity of CRISPR-Cas Systems

Detailed discussions on various CRISPR-Cas types and subtypes, such as Type I, Type II, and Type III systems, will be provided. We'll explore how each type functions and its unique features. Comparative genomic studies will be presented to highlight the diversity of CRISPR-Cas systems across different prokaryotic species. These data will demonstrate the evolutionary flexibility of these immune systems.

Objective 2: Mechanistic Insights into CRISPR Immunity

In Chapter 5, "The CRISPR Interference Stage," readers will gain a deep understanding of how CRISPR-Cas systems recognize and target invading genetic material. Molecular details of the Cascade complex and its role in interference will be elucidated. Structural biology findings, including high-resolution images of Cas proteins and complexes, will be included to provide concrete evidence of the mechanisms involved.

Objective 3: Evolutionary Dynamics of CRISPR Immunity

Chapter 7, "CRISPR Evolution and Horizontal Gene Transfer," will explore the coevolutionary arms race between phages and prokaryotes, showing how this has shaped CRISPR systems. Phylogenetic analyses and experimental studies demonstrating the dynamics of CRISPR array evolution and acquisition will be presented to support the discussions.

Objective 4: Practical Applications and Biotechnological Relevance

Chapter 8, "CRISPR-Cas Applications and Biotechnology," will delve into the transformative impact of CRISPR on genome editing and its potential in fields like agriculture, medicine, and biotechnology. Case studies will showcase real-world applications of CRISPR, including examples of genetically modified organisms (GMOs) and successful gene therapies.

Objective 5: Ethical and Regulatory Considerations

In Chapter 8, the discussion on "Ethical and Regulatory Considerations" will cover recent developments in bioethics and the regulation of CRISPR technologies, providing insights into the ongoing global debate on their use. Relevant legislation, international agreements, and societal opinions on CRISPR will be presented to underscore the importance of responsible innovation.

Objective 6: Future Prospects and Emerging Research Frontiers

In Chapter 19, "Future Directions in CRISPR Research," we'll explore emerging trends and unanswered questions in CRISPR research, including potential breakthroughs in the field. Cutting-edge studies and predictions from leading researchers will be cited, offering a glimpse into the future of CRISPR-Cas systems and their applications.

Interdisciplinary Nature

Understanding the mechanisms of CRISPR immunity in prokaryotes necessitates an interdisciplinary approach. This book will draw upon insights from molecular biology, genetics, bioinformatics, structural biology, ecology, and even ethics. By bridging these disciplines, readers will gain a holistic view of CRISPR-Cas systems and their impact on science, technology, and society.

Target Audience

The book is designed for a broad readership, including:

Researchers and Scientists: Those actively involved in CRISPR research will find detailed mechanistic insights and the latest developments in the field.

Educators and Students: Professors, teachers, and students seeking a comprehensive resource for courses in microbiology, genetics, molecular biology, and biotechnology will benefit from the foundational knowledge provided.

Policy Makers and Ethicists: Those interested in the ethical and regulatory aspects of CRISPR technology will gain valuable insights into the debates and considerations surrounding its use.

Science Enthusiasts: Anyone with an interest in cutting-edge science and its potential applications in areas like healthcare, agriculture, and conservation will find this book engaging and informative.

Structure of the Book

This book is organized into twenty chapters, each dedicated to a specific aspect of CRISPR-Cas immunity. From the fundamentals of prokaryotic genome defence to the ethical dilemmas surrounding its application, each chapter is meticulously crafted to provide a deep dive into the topic. Each chapter concludes with a summary and references for further exploration.

Hence, this book embarks on a journey to unravel the complexities of CRISPR immunity in prokaryotes. With a focus on mechanistic insights, evolutionary dynamics, and real-world applications, it aims to be a valuable resource for researchers, students, and enthusiasts alike, shedding light on one of the most fascinating biological phenomena of our time.

Chapter 2: Prokaryotic Genome Defence

2.1 The Vulnerability of Prokaryotic Genomes

Prokaryotic organisms, which include bacteria and archaea, are masters of adaptation and survival in diverse environments. Their genomes are compact, efficient, and highly responsive to environmental changes. However, these genomes are not immune to the constant threat of invasion by viruses, plasmids, and other mobile genetic elements. Understanding the vulnerability of prokaryotic genomes is essential to appreciate the significance of CRISPR-Cas systems in microbial defence.

Genomic Size and Vulnerability

Prokaryotic genomes are remarkably small compared to their eukaryotic counterparts. For instance, the model bacterium Escherichia coli boasts a genome of approximately 4.6 million base pairs, while the human genome contains over 3 billion base pairs. This size difference alone underscores the vulnerability of prokaryotic genomes. With fewer genetic resources at their disposal, any genetic disruption can have a profound impact.

Rapid Reproduction and Mutation Rates

Prokaryotes are known for their rapid reproduction rates, with some bacteria dividing every 20 minutes under optimal conditions. While this allows them to quickly adapt to changing environments, it also makes them susceptible to genetic changes introduced by invading elements. Mutations, both beneficial and detrimental, can arise frequently in bacterial populations due to their short generation times. This means that viruses and other invaders have a greater chance of finding genetic weaknesses within the bacterial population.

Lack of Protective Membrane-Bound Nuclei

Unlike eukaryotic cells, prokaryotic cells lack a membrane-bound nucleus. This absence means that prokaryotic genomes are constantly exposed to the cellular environment. Viruses and plasmids can directly access and interact with the bacterial genome without any physical barrier. This inherent vulnerability makes prokaryotic genomes prime targets for invasion.

The Arms Race: Coevolution with Invaders

Prokaryotes have engaged in an evolutionary arms race with their invaders for billions of years. Viruses, also known as bacteriophages, are a prominent example of these invaders. Bacteriophages are viruses that specifically infect bacteria. They have evolved sophisticated mechanisms to breach bacterial defences and hijack host machinery to replicate themselves.

In response, bacteria have developed various defence mechanisms to counter these invaders. One of the most notable defences is the restriction-modification system, which involves enzymes that recognize and degrade foreign DNA. However, bacteriophages have countered this by mutating their DNA to evade recognition. This constant back-and-forth escalation of genetic warfare highlights the ever-present vulnerability of prokaryotic genomes.

Horizontal Gene Transfer: A Double-Edged Sword

Horizontal gene transfer (HGT) is a fundamental process in prokaryotic evolution, allowing genes to move laterally between different organisms. While this process can provide prokaryotes with valuable genetic diversity and novel traits, it also exposes them to the risk of acquiring harmful genes, such as antibiotic resistance genes or virulence factors. HGT is often facilitated by mobile genetic elements like plasmids and transposons, which can act as vehicles for transferring genetic material.

Plasmids, for example, are small, circular pieces of DNA that can replicate independently within a bacterial cell. They often carry genes that provide selective advantages, such as antibiotic resistance genes. When plasmids are transferred horizontally to another bacterium, they confer the same advantages to the recipient cell. This process is a testament to the vulnerability of prokaryotic genomes to external genetic influences.

The Role of Mobile Genetic Elements

Mobile genetic elements (MGEs) are DNA sequences that have the ability to move within and between genomes. They include transposons, integrons, and insertion sequences, among others. MGEs can carry genes encoding for antibiotic resistance, toxins, and other adaptive traits, making them crucial players in the vulnerability of prokaryotic genomes.

Transposons, for instance, are DNA sequences capable of moving from one location in a genome to another. When they carry genes that confer resistance to antibiotics, they can spread these resistance genes throughout bacterial populations. This horizontal transfer of resistance genes poses a significant threat to human health by reducing the effectiveness of antibiotics.

Evidence from Genomic Analyses

Advancements in genomics have provided compelling evidence of the vulnerability of prokaryotic genomes. Large-scale sequencing projects have revealed the presence of numerous prophages (bacteriophage genomes integrated into bacterial genomes) in many bacterial genomes. These prophages are a testament to past encounters between bacteria and bacteriophages, where the bacterium survived and retained a piece of the phage genome as a genetic relic.

Furthermore, comparative genomics has shown that closely related bacterial strains can have significant variations in their genomic content. These variations often result from the acquisition or loss of genetic material through HGT events. For example, one strain of a pathogenic bacterium may possess genes that confer resistance to a specific antibiotic, while a closely related strain lacks these genes. This variation underscores how prokaryotic genomes are subject to constant genetic flux.

The Role of CRISPR-Cas Systems

It is within this context of genome vulnerability that CRISPR-Cas systems emerge as critical players in prokaryotic immunity. These systems provide prokaryotes with a remarkable adaptive defence mechanism against invasive genetic elements. By capturing and storing fragments of the genetic material of past invaders, prokaryotes can recognize and mount a rapid defence against future attacks.

In summary, prokaryotic genomes are vulnerable to invasion due to their small size, rapid reproduction and mutation rates, lack of protective membrane-bound nuclei, and the constant pressure of coevolution with invaders. Horizontal gene transfer and mobile genetic elements further exacerbate this vulnerability. The study of CRISPR-Cas systems is rooted in this vulnerability, as these systems represent a fascinating and sophisticated example of how prokaryotes have evolved to defend themselves in a dynamic and ever-changing genetic battleground.

2.2 CRISPR-Cas Systems as Genome Defence Mechanisms

In the vast microbial world, prokaryotes face an unceasing barrage of genetic invaders, from viruses to plasmids and other mobile genetic elements. To survive and thrive in this molecular battlefield, prokaryotes have developed an ingenious and highly effective defence mechanism known as CRISPR-Cas. In this section, we delve into the intricacies of how CRISPR-Cas systems function as genome defence mechanisms, using examples and data to illuminate their . significance.

The Arms Race with Invaders

Prokaryotic organisms, which include bacteria and archaea, inhabit diverse environments, each teeming with its own set of genetic invaders. Among these, bacteriophages (phages) are perhaps the most notorious. Phages are viruses that specifically infect bacteria and archaea, injecting their genetic material into host cells to replicate and produce progeny. To counter this constant threat, prokaryotes have evolved a variety of defence mechanisms, with CRISPR-Cas systems standing out as one of the most versatile and adaptable.

A Historical Glimpse

To understand the significance of CRISPR-Cas systems as genome defence mechanisms, it's essential to trace their historical roots. The term "CRISPR" stands for Clustered Regularly Interspaced Short Palindromic Repeats, which refers to the distinctive repeating sequences found in the genomes of many prokaryotes. These repeats were first noted in the late 1980s when Japanese scientists Yoshizumi Ishino and colleagues observed peculiar repetitions in the E. coli genome. However, it wasn't until 2002 that Francisco Mojica, a Spanish scientist, recognized their widespread presence in different prokaryotic species.

The groundbreaking insight came when Mojica noticed that these repeats were interspersed with unique sequences that he termed "spacers." These spacers, it turned out, were derived from genetic invaders, serving as a molecular memory of past encounters. This realization marked the inception of our understanding of CRISPR-Cas systems as adaptive immune systems in prokaryotes.

The CRISPR-Cas Arsenal

CRISPR-Cas systems can be thought of as an immune arsenal, with various types and subtypes akin to different weapon categories. Among the most well-studied is the Class 1 and Class 2 classification, each with its unique features. Class 1 systems are characterized by their multisubunit complexes, while Class 2 systems involve a single, large effector protein.

Data Insight 1: Class 1 systems, like the Type I and Type III CRISPR-Cas systems, are often found in complex with multiple Cas proteins, forming massive interference complexes. These systems are prevalent in environmental bacteria and archaea, underscoring their importance in diverse ecosystems.

Data Insight 2: Class 2 systems, exemplified by the Type II and Type V CRISPR-Cas systems, have gained significant attention in recent years due to their simplicity and ease of engineering for biotechnological applications. The Type II system, in particular, includes the well-known Cas9 protein, a centrepiece of modern genome editing.

The Intricate Dance of Immunity

To understand how CRISPR-Cas systems defend prokaryotic genomes, one must first grasp the intricate dance between the immune system and invaders. It begins with a process known as adaptation, during which the prokaryote captures and incorporates a piece of the invader's DNA – a spacer sequence – into its own CRISPR array. This is a remarkable feat, as the prokaryote must selectively choose spacers from the genetic material of the invader.

Example 1: A study conducted by Barrangou et al. in 2007 demonstrated this adaptation process in Streptococcus thermophilus, a bacterium used in the production of yogurt and cheese. The researchers found that S. thermophilus could selectively acquire spacers from phages that had infected it, providing a tangible example of how prokaryotes remember their attackers.

Data Insight 3: Recent genomic analyses have revealed the incredible diversity of spacers stored in prokaryotic genomes, showcasing the vast catalogue of invaders that CRISPR-Cas systems have encountered and remembered.

With the spacers integrated into the CRISPR array, the stage is set for the interference phase, where the prokaryote uses these spacers to recognize and target the invader's DNA for destruction. This recognition is highly specific, relying on base-pairing between the spacer sequence and the invader's genetic material, ensuring that only the targeted invader is neutralized.

Example 2: The specificity of CRISPR-Cas systems was exemplified in a study by Marraffini and Sontheimer in 2010, which investigated the Type III-A system in Staphylococcus epidermidis. They demonstrated that this system could effectively distinguish between closely related phages, emphasizing the precision of the prokaryotic immune response.

The Molecular Guillotine: Cas Proteins in Action

The effector machinery responsible for interference varies between CRISPR-Cas types. The well-known Cas9 protein, found in the Type II system, acts as a molecular guillotine, cleaving the invader's DNA at precise locations guided by the spacer's sequence.

Data Insight 4: The adaptability of the Cas9 system has spurred its use in genome editing. CRISPR-Cas9 technology has revolutionized genetics research and biotechnology, enabling targeted modifications in a wide range of organisms, including humans.

Other Cas proteins, like the Csm and Cmr complexes in Type III systems, function as RNA-guided complexes that not only target DNA but also degrade RNA. This dual functionality expands the scope of CRISPR-Cas immunity and highlights the diversity of strategies employed by prokaryotes to fend off invaders.

The Ongoing Evolution

One of the fascinating aspects of CRISPR-Cas systems is their constant evolution. Both prokaryotes and their invaders are engaged in an evolutionary arms race, each adapting to the strategies of the other.

Data Insight 5: Observations of CRISPR spacer dynamics have revealed a rapid turnover of spacers within prokaryotic populations, reflecting the perpetual need to adapt to new invaders. Additionally, phages have developed anti-CRISPR mechanisms to evade immunity, exemplifying the dynamic coevolution between prokaryotes and their adversaries.

CRISPR-Cas systems represent a remarkable example of how prokaryotes have harnessed the power of adaptive immunity to defend their genomes against relentless genetic invaders. From their historical discovery to their contemporary applications in biotechnology, these systems continue to captivate scientists and hold immense potential for the future, both in understanding microbial ecosystems and advancing genetic engineering.

2.3 How CRISPR-Cas Immunity Differs from Eukaryotic Immune Systems

The intricate world of immune defence mechanisms extends beyond the boundaries of eukaryotic cells into the realm of prokaryotes, where a unique and fascinating system called CRISPR-Cas (Clustered Regularly Interspaced Short Palindromic Repeats and CRISPR-associated proteins) provides an effective defence against invaders. While eukaryotic organisms, such as humans, employ an adaptive immune system based on antibodies and T cells, prokaryotes have evolved an entirely different strategy for protection. In this section, we will delve into the fundamental differences that set CRISPR-Cas immunity apart from eukaryotic immune systems, shedding light on the remarkable ways in which prokaryotes defend themselves against foreign genetic elements.

The Proactive vs. Reactive Nature of Immunity

One of the most striking distinctions between CRISPR-Cas immunity in prokaryotes and eukaryotic immune systems is their proactive versus reactive nature. Eukaryotic immune systems are inherently reactive; they respond to pathogens and invaders after they have breached the body's defences. In contrast, CRISPR-Cas immunity is remarkably proactive.

Example 1: Eukaryotic Immune System: Consider the human immune response to a viral infection. When a virus enters the body, it takes time for the immune system to recognize the invader, develop specific antibodies, and mount a defence. This delay can result in a period of vulnerability during which the virus can multiply and cause disease.

Example 2: CRISPR-Cas Immunity: In prokaryotes with CRISPR-Cas systems, the immune response is initiated even before an invader enters the cell. When a prokaryote encounters a new threat, it can rapidly acquire a piece of the invader's genetic material (a protospacer) and integrate it into its CRISPR array. This pre-emptive action allows the prokaryote to "remember" the invader and mount a swift and highly specific defence when the invader returns.

This proactive nature of CRISPR-Cas immunity is a remarkable adaptation that enables prokaryotes to defend themselves against a diverse array of invaders efficiently.

The Role of Memory in Immunity

Memory is a critical component of both eukaryotic and prokaryotic immune systems, but they achieve it through entirely different mechanisms.

Example 3: Eukaryotic Immune System: In the human immune system, memory is predominantly mediated by memory B cells and memory T cells. These cells "remember" previous infections, enabling a quicker and more effective response upon reinfection with the same pathogen.

Example 4: CRISPR-Cas Immunity: Prokaryotes, lacking specialized immune cells, rely on their CRISPR arrays to store genetic information about past invaders. When a prokaryote acquires a new spacer sequence, it effectively "remembers" the invader's genetic material. When the invader returns, the prokaryote can use the CRISPR-Cas system to precisely target and neutralize it.

The key distinction here is that while eukaryotic immune systems rely on memory cells, CRISPR-Cas systems use genetic memory in the form of spacer sequences.

Specificity and Precision

Another notable difference between CRISPR-Cas immunity and eukaryotic immune systems is the level of specificity and precision in recognizing and neutralizing threats.

Example 5: Eukaryotic Immune System: In eukaryotes, antibodies are produced to recognize broad classes of pathogens. Antibodies are highly specific but are generally not designed to recognize individual strains or variants of a pathogen. This specificity arises from the structure of antibodies' antigen-binding sites.

Example 6: CRISPR-Cas Immunity: CRISPR-Cas systems, on the other hand, exhibit an unparalleled level of specificity. When a prokaryote acquires a new spacer sequence, it is uniquely tailored to recognize a particular protospacer sequence in the invader's DNA or RNA. This high degree of specificity allows CRISPR-Cas systems to discriminate between closely related invaders, providing pinpoint accuracy in their defence.

Adaptive vs. Innate Immunity

Eukaryotic immune systems are typically categorized into innate and adaptive immunity. Innate immunity provides a rapid, nonspecific defence against a wide range of pathogens, while adaptive immunity offers a highly specific, memory-based response.

Example 7: Eukaryotic Immune System: The innate immune system in humans includes physical barriers like skin, as well as immune cells such as macrophages and neutrophils. These components act quickly but lack specificity.

Example 8: CRISPR-Cas Immunity: CRISPR-Cas immunity in prokaryotes combines aspects of both innate and adaptive immunity. It can provide a rapid response to previously encountered threats (adaptive) while maintaining a high degree of specificity (adaptive). This hybrid nature allows prokaryotes to efficiently defend against a wide range of invaders.

Molecular Machinery

Eukaryotic and prokaryotic immune systems also differ in the molecular machinery they employ for immune responses.

Example 9: Eukaryotic Immune System: In eukaryotes, immune responses involve a complex network of cells and signalling molecules. Antibodies, cytokines, and various immune cells like lymphocytes play crucial roles in detecting and neutralizing pathogens.

Example 10: CRISPR-Cas Immunity: Prokaryotic CRISPR-Cas immunity relies on a comparatively simpler molecular machinery. Cas proteins, guided by crRNAs, form the core of this system. This streamlined approach allows prokaryotes to mount a robust immune response with fewer components.

Inheritance of Immunity

The way immunity is inherited also sets these systems apart.

Example 11: Eukaryotic Immune System: In eukaryotes, immune memory is typically not inherited directly from one generation to the next. Each individual must build its own immune memory through exposure to pathogens.

Example 12: CRISPR-Cas Immunity: CRISPR-Cas immunity can be passed down from generation to generation through vertical gene transfer. When a prokaryote reproduces, its offspring inherit the same CRISPR arrays and spacer sequences, allowing them to be protected from the same threats.

Evolutionary Origins

Lastly, the evolutionary origins of these immune systems differ substantially.

Example 13: Eukaryotic Immune System: Eukaryotic immune systems, particularly adaptive immunity, are relatively recent in evolutionary terms. They likely emerged in jawed vertebrates around 500 million years ago.

Example 14: CRISPR-Cas Immunity: CRISPR-Cas immunity is ancient, with evidence suggesting its existence in prokaryotes for over a billion years. It has likely evolved alongside the ongoing battle between prokaryotes and their viral invaders.

CRISPR-Cas immunity in prokaryotes represents a remarkable departure from eukaryotic immune systems. Its proactive nature, reliance on genetic memory, specificity, and unique molecular machinery make it a powerful and efficient defence mechanism. Understanding these differences not only deepens our appreciation of the diversity of life but also informs the development of biotechnological applications, such as CRISPR-based genome editing, that have revolutionized the field of genetics and molecular biology.

Chapter 3: CRISPR-Cas System Components

3.1 Understanding the Core Components: Cas Proteins and CRISPR Arrays

In the intricate world of CRISPR-Cas systems, understanding the core components is fundamental to grasping how prokaryotes defend themselves against invaders. This chapter delves into the intricate machinery of CRISPR immunity, focusing on the two main protagonists: Cas proteins and CRISPR arrays. These elements form the backbone of the prokaryotic immune system, orchestrating a precise and highly adaptive response to invasive genetic material.

Cas Proteins: The Molecular Sentinels

Central to the CRISPR-Cas immune response are the Cas (CRISPR-associated) proteins. These multifunctional molecules act as molecular sentinels, executing the precise targeting and cleavage of invading DNA or RNA.

Cas proteins exhibit a remarkable diversity, with numerous subtypes identified across different prokaryotic species. Each subtype possesses distinct functions and mechanisms tailored to the specific challenges faced by their host organisms. For instance, the Cas9 protein, which gained immense fame in genome editing, is found in type II CRISPR systems, while type I systems utilize the Cas3 protein for target degradation.

Let's explore a notable example of Cas protein function – the Cas9 protein in the type II CRISPR-Cas system. Cas9 plays a pivotal role in recognizing and cleaving invading DNA. It relies on a guide RNA molecule, known as CRISPR RNA (crRNA), to identify its target. The crRNA is transcribed from the CRISPR array, where it retains a memory of past encounters with invaders.

The specificity of Cas9 in recognizing its target is governed by a short sequence called a protospacer adjacent motif (PAM). The PAM sequence acts as a molecular ZIP code, ensuring that Cas9 only engages with DNA that bears this specific marker. For example, in the Streptococcus pyogenes Cas9 system, the

PAM sequence is NGG (where N represents any nucleotide). This stringent requirement for a PAM sequence enhances the precision of the immune response.

Upon target recognition, Cas9 undergoes a conformational change, allowing it to form a double-stranded break in the invading DNA. This break is then repaired by the host cell's repair machinery, often introducing mutations or other genetic changes that render the invader nonfunctional. This mechanism exemplifies the exquisite specificity of Cas proteins, providing prokaryotes with a sophisticated defence against invasive genetic material.

CRISPR Arrays: The Genetic Memory Banks

In conjunction with Cas proteins, CRISPR arrays are the genetic memory banks that empower prokaryotes to recall past encounters with invaders. These arrays consist of short, repetitive DNA sequences interspersed with unique spacer sequences, which are derived from previous encounters with viruses or other genetic invaders.

Consider the CRISPR array as a prokaryotic "Most Wanted" list, where each spacer sequence represents a distinct genetic foe that the prokaryote has encountered and survived. The spacers serve as a record of past infections, allowing the prokaryote to recognize and mount a rapid response when faced with a familiar invader.

Let's examine a real-world example to illustrate the role of CRISPR arrays. In Streptococcus thermophilus, a bacterium used in yogurt fermentation, the CRISPR array contains spacers derived from bacteriophages that infect the bacterium. When a phage attempts to infect the bacterium, the CRISPR array's spacers are transcribed into crRNAs, which then guide Cas proteins to the corresponding protospacer sequences in the invading phage's DNA.

This targeted recognition and cleavage of the phage's DNA result in a swift immune response, preventing the phage from replicating and causing harm to the bacterium. The presence of the CRISPR array and its associated Cas proteins provides Streptococcus thermophilus with an effective defence mechanism that contributes to its survival during yogurt fermentation.

The diversity of CRISPR arrays is staggering, with varying numbers of repeats and spacers found in different prokaryotic species. This diversity reflects the prokaryote's unique evolutionary history and the array's adaptation to the types of invaders it encounters in its ecological niche.

Beyond their role in immunity, CRISPR arrays are dynamic elements that evolve over time. New spacers can be added through a process called spacer acquisition, which occurs when the prokaryote captures a snippet of the invader's genetic material and incorporates it into its CRISPR array. This process provides the prokaryote with an expanded repertoire of spacers, enhancing its ability to fend off future attacks.

In the complex dance of prokaryotic immunity, Cas proteins and CRISPR arrays choreograph the defence against invading genetic material. Cas proteins act as the vigilant guardians, ensuring that only the genetic material bearing the correct PAM sequence is targeted and cleaved. CRISPR arrays, on the other hand, function as the genetic memory banks, storing a record of past encounters and allowing the prokaryote to respond swiftly to familiar invaders.

Understanding the interplay between these core components is essential to comprehend the precision and adaptability of CRISPR immunity. In the subsequent chapters, we will explore the intricacies of how these components work together and dive deeper into the diverse world of CRISPR-Cas systems found in prokaryotes, each with its unique mechanisms and evolutionary history.

3.2 Types and Subtypes of CRISPR-Cas Systems

The CRISPR-Cas system is a remarkably diverse and adaptive immune system found in prokaryotic organisms. Since its discovery, scientists have identified several types and subtypes of CRISPR-Cas systems, each with its own unique features and mechanisms. This diversity is a testament to the evolutionary arms race between prokaryotes and their viral invaders. In this section, we will delve into the various types and subtypes of CRISPR-Cas systems and explore their distinctive characteristics.

Type I CRISPR-Cas Systems

Type I CRISPR-Cas systems are one of the most prevalent and well-studied types. They are characterized by the presence of a multisubunit complex known as Cascade (CRISPR-associated complex for antiviral defence) and the signature Cas3 protein.

Cascade Complex: Cascade is a large, multi-protein complex that plays a central role in the interference stage of Type I systems. It is composed of multiple Cas proteins and a CRISPR RNA (crRNA) molecule. The crRNA guides Cascade to the target DNA by base-pairing with a complementary sequence, leading to the recruitment of Cas3 for DNA degradation.

Cas3 Nuclease: Cas3, a hallmark of Type I systems, is an ATP-dependent helicase-nuclease. It unwinds and degrades the invading DNA, effectively silencing the threat. The concerted action of Cascade and Cas3 makes Type I systems highly effective at neutralizing foreign DNA.

Example: A study by Jackson et al. (2017) revealed the crystal structure of a Type I-E Cascade complex bound to its crRNA and target DNA, providing critical insights into the molecular mechanisms underlying target recognition and DNA interference in Type I systems.

Type II CRISPR-Cas Systems

Type II systems are exemplified by the widely-used CRISPR-Cas9 system, which has revolutionized genome editing. Type II systems are relatively simple compared to other types and rely on a single Cas protein for interference.

Cas9 Nuclease: Cas9 is an RNA-guided endonuclease responsible for target recognition and cleavage. It uses a synthetic guide RNA (sgRNA) to locate the target DNA sequence, allowing for precise and programmable genome editing.

Example: The pioneering work of Doudna and Charpentier in 2012 demonstrated the adaptability of the Type II CRISPR-Cas system for genome editing in a wide range of organisms, showcasing its potential for biotechnological applications.

Type III CRISPR-Cas Systems

Type III systems are unique in that they combine both interference and adaptation machinery in a single multi-subunit complex known as Csm (CRISPR-associated complex for antiviral defence). This feature blurs the line between adaptation and interference and suggests an intriguing connection between the two processes.

Csm Complex: The Csm complex comprises multiple Cas proteins and a crRNA similar to Type I systems. It targets invading RNA rather than DNA, cleaving it to thwart viral replication. Additionally, Type III systems may be involved in priming the adaptation process by generating new spacers from invader RNA.

Example: A study by Tamulaitis et al. (2018) demonstrated that the Type III-A Csm complex is involved in both crRNA-guided RNA cleavage and primed adaptation, shedding light on the complex interplay between these processes.

Type IV CRISPR-Cas Systems

Type IV systems are one of the latest additions to the CRISPR-Cas family. They are defined by a distinct signature protein, Csm4, which lacks the interference activities found in other types.

Csm4 Protein: Csm4 is unique to Type IV systems and is not directly involved in interference. Instead, it is thought to play a regulatory role, possibly modulating the activity of neighbouring Type III systems. This suggests a complex network of CRISPR-Cas systems within prokaryotes.

Example: Recent metagenomic studies have identified Type IV systems in diverse environments, highlighting their prevalence and potential ecological significance in microbial communities.

Type V CRISPR-Cas Systems (Cpf1/Cas12)

Type V systems, represented by Cpf1 (or Cas12), are another exciting addition to the CRISPR-Cas toolkit. Cpf1 exhibits distinct features compared to Cas9 in terms of target recognition and cleavage.

Cpf1 Nuclease: Cpf1 is an endonuclease that recognizes a T-rich protospacer adjacent motif (PAM) and generates staggered DNA breaks. This unique PAM requirement expands the targetable sequences, making Cpf1 a valuable tool for genome editing.

Example: A study by Zetsche et al. (2015) introduced Cpf1 as an alternative to Cas9 for genome editing, showcasing its potential for precise and versatile genetic modifications.

Type VI CRISPR-Cas Systems (C2c2/Cas13)

Type VI systems, represented by C2c2 (or Cas13), are known for their RNA-targeting capabilities. These systems have gained attention for their potential applications in RNA manipulation and diagnostics.

C2c2 Ribonuclease: C2c2 is an RNA-guided ribonuclease that targets and cleaves RNA molecules. It has been harnessed for its RNA interference properties, making it a promising tool for RNA-based applications.

Example: Abudayyeh et al. (2016) demonstrated the use of C2c2/Cas13 for programmable RNA cleavage, opening new avenues for RNA-focused research and diagnostics.

The diversity of CRISPR-Cas systems is a testament to the intricate strategies prokaryotes employ to defend against viral invaders. Each type and subtype comes with its own set of proteins and mechanisms, making them adaptable to different environmental challenges. This diversity not only enriches our understanding of prokaryotic immunity but also expands the toolbox for biotechnological applications. As research in this field continues to advance, we can expect even more exciting discoveries and innovations to emerge from the world of CRISPR-Cas systems.

3.3 Accessory Proteins and Their Roles in Immunity

In our exploration of CRISPR-Cas systems and their mechanisms, we have examined the core components, including Cas proteins and CRISPR arrays. However, the intricate world of prokaryotic immunity is not solely reliant on these primary elements. Accessory proteins play a crucial role in the functioning and regulation of CRISPR-Cas systems. In this section, we delve into the fascinating realm of accessory proteins and their multifaceted roles in the prokaryotic immune response.

In our exploration of CRISPR-Cas systems and their mechanisms, we have examined the core components, including Cas proteins and CRISPR arrays. However, the intricate world of prokaryotic immunity is not solely reliant on these primary elements. Accessory proteins play a crucial role in the functioning and regulation of CRISPR-Cas systems. In this section, we delve into the fascinating realm of accessory proteins and their multifaceted roles in the prokaryotic immune response.

While Cas proteins and CRISPR arrays form the backbone of the CRISPR-Cas system, accessory proteins are the supporting actors that fine-tune and enhance the system's efficiency. These proteins are often less conserved across different CRISPR-Cas types and subtypes, reflecting the diversity of mechanisms and strategies that prokaryotes employ to fend off invading genetic elements.

Cascade Complex: The RNA-Guided Surveillance Team

One of the key accessory proteins in the CRISPR-Cas system is the Cascade complex, which plays a pivotal role in the interference stage. Cascade stands for CRISPR-associated complex for antiviral defense and is aptly named for its function in searching for and binding to the target DNA.

Cascade is composed of multiple proteins, with CasA and CasB forming the backbone of the complex and CasC, CasD, CasE, and CasF forming a surveillance complex that associates with a mature crRNA molecule. The Cascade complex serves as the "search and destroy" team, actively seeking out the complementary DNA sequence to the crRNA.

Example: Cascade Surveillance in Action

Imagine a scenario in which a bacterium harbours a CRISPR-Cas system with a preloaded crRNA sequence that matches a specific viral DNA. When the bacterium encounters this virus, the Cascade complex is activated. The crRNA, guided by its sequence complementarity, pairs with the viral DNA, initiating a molecular match akin to fitting puzzle pieces together.

Once the Cascade complex has localized the viral DNA, it recruits other Cas proteins, such as Cas3, to execute the destructive phase of interference. This involves the degradation of the invading DNA, effectively neutralizing the threat. Thus, the Cascade complex acts as the sentinel, identifying intruders and marking them for elimination.

Csm and Cmr Complexes: The Effector Arsenal

In addition to Cascade, prokaryotes possess other effector complexes known as Csm (Type II CRISPR-Cas systems) and Cmr (Type III CRISPR-Cas systems). These complexes are responsible for carrying out the destructive phase of immunity, once the target DNA has been identified.

Example: Csm and Cmr Complexes in Immunity

Let's consider a bacterium equipped with a Type III CRISPR-Cas system, housing the Cmr complex. Upon encountering a viral invader, this complex is activated. Cmr complexes are distinctive because they can target RNA molecules, not just DNA. In this case, the crRNA guides the Cmr complex to the viral RNA, and Cmr proteins proceed to cleave and degrade it.

Similarly, Type II CRISPR-Cas systems possess the Csm complex, which follows a similar principle but targets DNA instead of RNA. These effector complexes, along with Cascade, constitute the prokaryotic immune system's weaponry, effectively dismantling the genetic material of the invader.

Csy Complex: The Interference Engine in Type I Systems

Type I CRISPR-Cas systems, characterized by their multi-subunit interference complexes, employ the Csy complex to execute interference. The Csy complex comprises Cas proteins that work together to target and cleave invading DNA.

Example: Csy Complex Functionality

Consider a scenario where a bacterium utilizes a Type I CRISPR-Cas system with a Csy complex. Upon viral infection, the CRISPR array is transcribed into a long precursor CRISPR RNA (pre-crRNA), which is then processed into mature crRNAs. These crRNAs guide the Csy complex to the invading DNA.

The Csy complex scans for a PAM (Protospacer Adjacent Motif) sequence adjacent to the target DNA, ensuring that it is the correct genetic material to be cleaved. Once the PAM is recognized, the Csy complex initiates the destruction of the target DNA, thereby neutralizing the threat. The precision of this process, guided by the crRNA, is a testament to the remarkable adaptability of CRISPR-Cas systems.

Csm6 and Csx1: The Guardians of Self-Recognition

Self-recognition is a crucial aspect of CRISPR-Cas immunity. Prokaryotes must avoid targeting their own genetic material inadvertently. To prevent this, accessory proteins such as Csm6 and Csx1 have evolved.

Example: Self-Recognition Mechanisms

In a bacterial cell, Csm6 monitors the cell's own transcriptome. It is equipped with RNase activity and can degrade the cell's own RNA molecules. When foreign RNA, such as viral RNA, is present in the cell, Csm6 is inactivated, allowing it to accumulate and signal the presence of an invader.

Similarly, Csx1 acts as a guardian against self-targeting in Type III CRISPR-Cas systems. It monitors the cell's own transcripts and, like Csm6, becomes active in the presence of foreign RNA, thereby distinguishing self from non-self.

C2c2/Cas13: RNA-Targeting in Class 2 Systems

Class 2 CRISPR-Cas systems are characterized by single, multidomain effector proteins such as C2c2 (now known as Cas13). These proteins have gained prominence for their RNA-targeting capabilities.

Example: Cas13 in Action

Let's envision a Class 2 CRISPR-Cas system within a bacterium. Upon viral infection, the Cas13 protein is guided by the crRNA to the viral RNA. Unlike DNA-targeting systems, Cas13 doesn't cleave the target RNA but rather becomes activated and initiates a collateral cleavage of nearby RNA molecules. This indiscriminate cleavage effectively disrupts the cellular processes of the invader.

The versatility of Class 2 systems, particularly Cas13, has led to their exploration in various biotechnological applications, including RNA interference and diagnostic tools.

Csm4 and Csx3: Enhancers of Interference

Accessory proteins like Csm4 and Csx3 function as enhancers of interference in Type I systems. They assist in the efficient binding and cleavage of target DNA.

Example: Enhanced Interference

In a Type I CRISPR-Cas system, Csm4 stabilizes the Csm complex, ensuring its integrity during target recognition and cleavage. Csx3, on the other hand, functions as an auxiliary protein, facilitating target DNA binding and subsequent degradation.

These accessory proteins work in concert with the core components of the CRISPR-Cas system to bolster its efficiency in combating invaders.

The Ensemble of Accessory Proteins

In the grand symphony of CRISPR-Cas immunity, accessory proteins are the supporting musicians, harmonizing with Cas proteins and crRNAs to create a formidable defence against invasive genetic elements. From the sentinel function of the Cascade complex to the effector roles of Csm, Cmr, and Csy complexes, these proteins showcase the intricate and adaptable nature of prokaryotic immunity.

Furthermore, self-recognition mechanisms, exemplified by Csm6, Csx1, and others, demonstrate the sophistication of these systems in distinguishing self from non-self. Class 2 systems, represented by Cas13, expand the repertoire of RNA-targeting strategies.

As we continue our exploration of CRISPR-Cas systems, it becomes evident that understanding the roles of accessory proteins is crucial not only for deciphering the mechanisms of prokaryotic immunity but also for harnessing the power of CRISPR in biotechnology and medicine. These proteins are the unsung heroes in the battle against invaders, ensuring the survival and evolution of prokaryotic life.

Chapter 4: The CRISPR Adaptation Process

4.1 Spacer Acquisition: How Prokaryotes Remember Invaders

Spacer acquisition is a fundamental process in the CRISPR-Cas immune system, allowing prokaryotes to remember and defend against past invaders. This subsection delves into the mechanisms and significance of spacer acquisition, providing examples and data to illustrate its importance in prokaryotic immunity.

Spacer Acquisition Mechanisms

Prokaryotes acquire new spacers through a process called "primed adaptation." This process begins when a prokaryotic cell survives an encounter with a new phage or plasmid. The cell captures a short piece of the invader's genetic material, known as a "protospacer," and integrates it into the CRISPR array as a new spacer. Here's how it works:

Protospacer Recognition: Upon phage or plasmid invasion, the prokaryotic cell's defence mechanisms, such as restriction-modification systems and innate immunity, may fend off the invader partially. This sets the stage for CRISPR-Cas to come into play. The Cascade complex, a central player in the CRISPR immune system, helps in the initial recognition of the protospacer sequence within the invader's genetic material.

Example 1: Cascade Complex Recognition: Studies using the model bacterium Escherichia coli (E. coli) have shown that the Cascade complex is essential for the initial recognition of protospacer sequences. In a study published in *Nature*, researchers observed that mutants lacking the Cascade complex were unable to acquire new spacers efficiently.

Protospacer Capture: Once the Cascade complex recognizes the protospacer, it forms a complex with another key protein, Cas1-Cas2. This complex helps in capturing the protospacer sequence.

Example 2: Cas1-Cas2 Complex: In a groundbreaking study published in *Science* [2], researchers elucidated the crystal structure of the Cas1-Cas2 complex. This structure revealed how Cas1-Cas2 functions as a molecular ruler, ensuring that the captured protospacer is the correct size for integration.

Integration into the CRISPR Array: The captured protospacer is then integrated into the CRISPR array, which is essentially a library of past invaders. The Cas1-Cas2 complex, in conjunction with other accessory proteins, facilitates the integration process.

Example 3: Spacer Integration Efficiency: Research conducted on Streptococcus thermophilus demonstrated that the efficiency of spacer integration is influenced by various factors, including the availability of Cas1-Cas2 complexes and the sequence similarity between the protospacer and existing spacers in the array.

Significance of Spacer Acquisition

Spacer acquisition is not merely a molecular process; it has profound implications for prokaryotic immunity and microbial evolution. Here are some key reasons why spacer acquisition is essential:

Adaptive Immunity: Spacer acquisition is the core of the CRISPR-Cas adaptive immune system. By capturing and storing genetic information from invaders, prokaryotes can recognize and mount a defence against the same or closely related invaders in the future.

Example 4: Adaptive Immunity in Streptococcus pyogenes: A study in *Nature* highlighted how Streptococcus pyogenes, a pathogenic bacterium, uses CRISPR-Cas adaptive immunity to protect itself from recurrent phage attacks. Spacer acquisition allows it to rapidly adapt to new phage threats.

Diversity in CRISPR Arrays: Spacer acquisition contributes to the diversity of spacers in the CRISPR array.

This diversity is crucial because it increases the chances of having a spacer that matches a future invader.

Example 5: Spacer Diversity in Haloferax volcanii: Research on Haloferax volcanii revealed that this archaeon has a highly diverse CRISPR array, with hundreds of unique spacers. This diversity enhances its immunity against a wide range of viruses.

Evasion of Phage Countermeasures: Phages, the viruses that infect bacteria, have evolved various countermeasures to evade CRISPR-Cas immunity. Spacer acquisition allows prokaryotes to keep pace with evolving phages.

Example 6: Phage Evolution and Spacer Acquisition: A study published in *Cell* documented how phages can rapidly evolve to escape CRISPR immunity. In response, prokaryotes continuously acquire new spacers to counter these evolving phages.

Long-Term Memory: The spacers integrated into the CRISPR array serve as a long-term memory of past encounters. This memory can last for generations, ensuring that the prokaryotic population remains prepared for recurring threats.

Example 7: Long-Term Memory in Bacillus cereus: Bacillus cereus, a soil bacterium, was found to retain spacers acquired from phage infections for several generations. This long-term memory is a testament to the durability of CRISPR immunity.

Spacer Acquisition and Evolutionary Dynamics

Spacer acquisition is intimately linked to the evolutionary dynamics of prokaryotes and their adversaries. Understanding these dynamics provides insights into how CRISPR-Cas systems have shaped microbial ecosystems.

Arms Race Between Prokaryotes and Phages: Spacer acquisition is a crucial element in the coevolutionary arms race between prokaryotes and phages. As prokaryotes acquire new spacers to defend against phages, phages evolve to evade these defences, leading to an ongoing cycle of adaptation and counter-adaptation.

Example 8: Coevolution in Pseudomonas aeruginosa: Research on Pseudomonas aeruginosa and its phage adversaries demonstrated the rapid coevolution of both parties. Spacer acquisition by P. aeruginosa was a key factor in this dynamic.

Spacer Acquisition as a Driver of Diversity: The continuous acquisition of spacers contributes to the genetic diversity within prokaryotic populations. This diversity can have broader ecological implications, such as influencing microbial community structure.

Example 9: Microbial Community Dynamics: Studies in environmental microbiology have shown that spacer acquisition can influence the composition of microbial communities. In a study published in *Environmental*

Microbiology Reports, researchers observed shifts in community composition driven by CRISPR-Cas activity.

Horizontal Gene Transfer and Spacer Acquisition: Spacer acquisition can also facilitate horizontal gene transfer (HGT) between prokaryotes. When a prokaryote acquires a spacer from a plasmid, it may gain resistance to the plasmid's genes, potentially influencing the spread of antibiotic resistance genes.

Example 10: Spacer-Mediated HGT: Investigations into the role of spacer acquisition in HGT, as seen in *mBio*, have revealed that spacers can influence the exchange of genetic material between prokaryotes, with implications for antibiotic resistance dissemination.

Spacer acquisition is a multifaceted process that lies at the heart of CRISPR-Cas adaptive immunity in prokaryotes. It not only provides a mechanism for defending against invaders but also shapes the evolutionary dynamics of prokaryotic populations and their interactions with phages and plasmids. This process exemplifies the elegance and complexity of nature's immune systems, offering insights into how prokaryotes remember and respond to their ever-evolving microbial adversaries.

4.2 Mechanisms of Protospacer Selection and Integration

In the intricate world of CRISPR-Cas immunity, the process of protospacer selection and integration is a fundamental step that allows prokaryotic organisms to memorize and defend against invasive genetic elements. This subsection delves into the mechanisms behind this process, shedding light on the fascinating ways prokaryotes discern friend from foe and incorporate foreign DNA fragments into their CRISPR arrays.

Protospacer Selection: The Discriminating Eye

Protospacers, which are short segments of DNA derived from invasive elements such as viruses or plasmids, serve as the memory bank of CRISPR-Cas systems. However, prokaryotes must be judicious in their choice of protospacers to ensure that only genetic material from invaders is integrated. Let's explore the mechanisms they employ for this critical selection process.

PAM Recognition: One of the first checkpoints in protospacer selection is the presence of a Protospacer Adjacent Motif (PAM). PAMs are short, conserved sequences adjacent to the protospacers that are recognized by Cas proteins. For example, the PAM sequence 'NGG' is recognized by the widely-used Cas9 nuclease in type II CRISPR-Cas systems found in many bacteria, including Streptococcus pyogenes. This recognition mechanism ensures that only foreign DNA with the appropriate PAM sequence is targeted, preventing accidental self-destruction.

Example: In Streptococcus pyogenes, the Cas9 protein identifies the PAM sequence 'NGG' in the target DNA, such as a viral genome. When it encounters this PAM, it proceeds with protospacer binding and interference.

Protospacer Length and Sequence Matching: Prokaryotes also scrutinize the length and sequence of the protospacer for a precise match with the target DNA. The level of sequence identity required for interference can vary among different CRISPR-Cas systems. Some systems demand near-identical matches, while others tolerate mismatches to a certain degree.

Example: In the type I-F CRISPR-Cas system found in Pseudomonas aeruginosa, a protospacer with a 6- to 8-nucleotide match to the target DNA is usually sufficient for interference. This flexibility allows the system to adapt to a broader range of invaders.

Spacer Selection and Avoidance of Self-Targeting: The selection of appropriate spacers is crucial. To avoid self-targeting, prokaryotes employ various mechanisms to differentiate their own DNA from foreign DNA. One such mechanism is the differential methylation of self and non-self DNA. The host organism methylates its own DNA, making it less susceptible to CRISPR interference.

Example: In Escherichia coli, self-DNA is often methylated, whereas DNA from invading phages lacks this methylation. The CRISPR system can distinguish between these types of

DNA and select non-methylated DNA as a suitable protospacer.

Adaptation Interference: In some cases, CRISPR-Cas systems exhibit a phenomenon called "adaptation interference." This occurs when a new spacer is added to the CRISPR array that matches a region within the host genome. To prevent autoimmunity, the system employs mechanisms like anti-CRISPR proteins or simply avoids adding self-targeting spacers.

Example: The type II-A CRISPR-Cas system in Francisella novicida employs the anti-CRISPR protein AcrF1 to suppress the acquisition of spacers that match the host genome. This ensures that the system does not target its own DNA inadvertently.

Protospacer Integration: Building the Immune Archive

Once a suitable protospacer is selected, the prokaryotic organism must integrate it into its CRISPR array to maintain a record of the invader. This process involves several intricate steps.

Cas1-Cas2 Complex: The Cas1-Cas2 complex is central to protospacer integration. These proteins work together to capture the selected protospacer and integrate it into the CRISPR array as a new spacer. Cas1 is responsible for cleaving the protospacer from the invader's DNA, while Cas2 aids in integration into the CRISPR array.

Example: In Streptococcus thermophilus, the Cas1-Cas2 complex is responsible for integrating new spacers into the CRISPR array, ensuring a continually updated archive of invaders.

Leader Sequence Recognition: Prokaryotes often have a leader sequence upstream of the CRISPR array. This sequence plays a vital role in the integration process, as it guides the Cas1-Cas2 complex to the correct location for spacer integration.

Example: In Sulfolobus solfataricus, a crenarchaeon with a type I-A CRISPR-Cas system, the leader sequence directs the Cas1-Cas2 complex to integrate spacers at the correct end of the CRISPR array.

Spacer Orientation: The orientation of the integrated spacer is crucial. It determines which strand of the invader's DNA is targeted during interference. Proper orientation ensures effective immunity against future encounters with the same invader.

Example: In Haloarcula hispanica, a halophilic archaeon with a type I-B CRISPR-Cas system, spacers are integrated in a specific orientation to target the correct strand of the invader's DNA.

Spacer Location in the Array: The location of the integrated spacer within the CRISPR array can affect the timing and efficiency of interference. In some systems, newly acquired spacers are integrated at the leader end, while in others, they are added to the opposite end.

Example: In Streptococcus mutans, a bacterium with a type II-C CRISPR-Cas system, newly acquired spacers are integrated at the leader end, allowing for efficient targeting of invaders.

The mechanisms of protospacer selection and integration in CRISPR-Cas systems are a testament to the remarkable precision and adaptability of prokaryotic immune systems. These processes ensure that prokaryotes can distinguish between self and non-self DNA and build an ever-expanding archive of invaders. As our understanding of CRISPR-Cas systems deepens, it opens up exciting possibilities for harnessing these mechanisms in biotechnology and medicine, further underscoring the importance of ongoing research in this field.

4.3 Implications for Diversity in CRISPR Arrays

CRISPR arrays, the DNA archives of past encounters with mobile genetic elements like viruses and plasmids, are a testament to the remarkable diversity of life on Earth. These arrays are at the heart of the CRISPR-Cas adaptive immune system in prokaryotes, serving as libraries of genetic memory, enabling the cell to recognize and defend against specific invaders. In this subsection, we delve into the fascinating world of CRISPR array diversity, exploring how these arrays accumulate and evolve, and the implications of this diversity for prokaryotic immunity.

The Dynamic Nature of CRISPR Arrays

CRISPR arrays are dynamic structures that continuously change and adapt in response to new threats. The key process governing this change is spacer acquisition, where prokaryotic organisms incorporate short DNA fragments from invading elements into their CRISPR array. This incorporation is a critical aspect of the CRISPR adaptive immune response. It's like adding a new book to the library, where each book represents a different invader.

Spacer acquisition can occur in two primary ways: naïve adaptation and primed adaptation. In naïve adaptation, the cell encounters a novel invader, and if it survives, it may incorporate a piece of the invader's DNA as a spacer into the CRISPR array. In primed adaptation, the cell already has a spacer that is somewhat similar to the invader's DNA. When it encounters a similar invader, it can efficiently integrate spacers into the array.

Diversity Through Naïve Adaptation

Naïve adaptation contributes significantly to the diversity of CRISPR arrays. When a prokaryotic cell encounters a novel virus or plasmid, it may incorporate a fragment of the invader's DNA into its CRISPR array as a new spacer. This sequence is typically taken from the protospacer region of the invader's DNA, which is the target of the CRISPR-Cas interference machinery.

For instance, consider a bacterium exposed to a new phage (virus) for the first time. If the bacterium survives the infection, it can acquire a spacer from the phage's DNA and integrate it into its CRISPR array. This process allows the bacterium's descendants to recognize and mount a more effective defence against the same phage in the future.

The diversity generated through naïve adaptation is staggering. Different prokaryotic populations exposed to different invaders accumulate unique sets of spacers in their CRISPR arrays over generations. This diversity in spacer sequences is akin to a species developing a diverse array of antibodies to combat various pathogens. It enhances the ability of a prokaryotic community to collectively defend against a wide range of invaders.

Primed Adaptation: Efficiently Building Diversity

Primed adaptation is a mechanism by which prokaryotes can efficiently build diversity in their CRISPR arrays when encountering similar invaders. This mechanism relies on the recognition of a "seed" or "priming" spacer that partially matches the invader's DNA. When the priming spacer recognizes a related invader, the CRISPR-Cas system is primed to incorporate spacers efficiently.

Imagine a scenario where a bacterium has already acquired a spacer that partially matches a phage's DNA. When this bacterium encounters a closely related phage, the priming spacer can recognize and bind to the phage's DNA more efficiently, leading to the rapid incorporation of new spacers into the CRISPR array. This mechanism enables prokaryotes to quickly adapt to evolving invaders and build a diverse repertoire of spacers targeting related threats.

Spacer Diversity as a Measure of Immune History

The diversity of spacers in a CRISPR array serves as a historical record of a prokaryotic organism's encounters with invaders. Scientists can study these arrays to gain insights into the types of threats an organism has faced throughout its evolutionary history.

For example, by analysing the spacers in the CRISPR arrays of a bacterial population from a particular environment, researchers can discern the prevalence of specific viruses or plasmids in that environment. This information can be invaluable for understanding the dynamics of host-virus interactions in complex microbial ecosystems.

Implications for Immune Memory and Specificity

The diversity in CRISPR arrays has profound implications for the immune memory and specificity of prokaryotic organisms. Each spacer represents a specific encounter with an invader, and the more diverse the spacers, the broader the range of invaders a prokaryote can recognize and defend against.

Consider a bacterium with a CRISPR array containing spacers from ten different phages. This bacterium has a broader immune memory than one with spacers from only two phages. It can recognize and mount an immune response against a more extensive spectrum of threats. This diversity in immune memory enhances the adaptability and resilience of prokaryotic populations in the face of ever-evolving invaders.

Moreover, the specificity of CRISPR immunity is rooted in spacer diversity. Each spacer is specific to a particular sequence in the invader's DNA, ensuring that the immune response is precisely targeted. This specificity minimizes the risk of self-attack, where the CRISPR-Cas system accidentally targets the host's DNA.

Spacer Diversity in Natural Populations

The diversity in CRISPR arrays isn't limited to laboratory studies; it's a common feature in natural prokaryotic populations. For example, a study of marine bacteria found that their CRISPR arrays contained spacers targeting a wide variety of phages and plasmids commonly found in the marine environment. This diversity reflected the constant arms race between bacteria and their viral adversaries in the oceans.

Similarly, studies in soil bacteria have revealed diverse CRISPR arrays with spacers targeting various plasmids and viruses commonly found in soil ecosystems. These findings underscore the importance of CRISPR diversity in natural settings, where prokaryotes must defend against a multitude of invaders to survive and thrive.

Evolutionary Implications of CRISPR Array Diversity

The diversity of CRISPR arrays has profound evolutionary implications. It not only enhances the survival of individual prokaryotic organisms but also contributes to the overall fitness of prokaryotic populations. Over time, the accumulation of diverse spacers in CRISPR arrays can lead to the emergence of new immune strategies and the evolution of more efficient immune systems.

Furthermore, the coevolutionary dynamics between prokaryotes and their invaders are shaped by the diversity of CRISPR arrays. As invaders evolve to evade existing immune responses, prokaryotes with more diverse CRISPR arrays have a better chance of recognizing and defending against these evolving threats. This ongoing arms race drives the continual adaptation and diversification of both prokaryotes and their invaders.

The diversity of CRISPR arrays is a testament to the remarkable adaptability of prokaryotic organisms. These arrays, constantly changing and evolving through spacer acquisition, enable prokaryotes to recognize and defend against a wide range of invaders. The historical record encoded in CRISPR arrays provides insights into the immune history of prokaryotic populations, while the specificity and efficiency of CRISPR immunity are rooted in this diversity. This diversity not only enhances the survival of individual organisms but also drives the evolution of more robust and effective immune systems, shaping the coevolutionary dynamics between prokaryotes and their invaders in the ever-changing microbial world.

Chapter 5: The CRISPR Interference Stage

5.1 Target Recognition and Protospacer PAM Sequences

In the intricate world of CRISPR immunity, the ability to accurately identify and target specific invader DNA sequences is paramount. This process hinges on the recognition of Protospacer Adjacent Motifs (PAMs) within the target DNA, a crucial step in CRISPR interference. In this section, we delve into the fascinating mechanisms behind target recognition, explore the significance of PAM sequences, and examine the role they play in ensuring the precision of CRISPR immunity.

The Molecular Precision of CRISPR-Cas Targeting

At the heart of CRISPR immunity lies the remarkable precision with which prokaryotic cells recognize and cleave foreign DNA while sparing their own genomic material. This precision begins with the acquisition of short DNA sequences from invading viruses or plasmids, known as protospacers, into the CRISPR array. However, it is the subsequent targeting of these protospacers within the invaders' genetic material that distinguishes CRISPR-Cas systems as highly sophisticated immune mechanisms.

Protospacer Adjacent Motifs (PAMs): The Recognition Code

To accomplish this precision targeting, CRISPR-Cas systems rely on Protospacer Adjacent Motifs or PAMs. PAMs are short, conserved DNA sequences adjacent to protospacers and are a hallmark of CRISPR interference. The presence of a PAM sequence is a critical determinant for whether a particular protospacer will be targeted by the Cas protein machinery.

5.3 The Significance of PAMs

PAMs Prevent Self-Targeting

One of the central functions of PAM sequences is to prevent self-targeting. Since the CRISPR array contains protospacers derived from previous invaders, a strict self-versus-nonself discrimination is crucial. PAM sequences act as a protective shield for the host genome. They ensure that the Cas proteins only bind to and cleave DNA that possesses the correct PAM sequence, which is typically absent from the host genome. This discrimination prevents the CRISPR-Cas system from inadvertently targeting and damaging its own genetic material.

PAM Diversity Reflects Genetic Variation

PAM sequences are not universal among all CRISPR-Cas systems but are specific to each subtype. This diversity in PAM sequences reflects the genetic variation among invading phages and plasmids. Different phages and plasmids may carry distinct PAM sequences, and CRISPR-Cas systems have evolved to recognize the PAM sequences of the invaders they encounter most frequently. This adaptability allows CRISPR-

Cas systems to remain effective against a wide array of invaders.

The Role of PAM Recognition in Interference

Once a protospacer has been incorporated into the CRISPR array and the Cas proteins are primed for action, the PAM sequence plays a pivotal role in the interference stage of CRISPR immunity.

PAM Recognition Initiates Interference

When the Cas complex scans the cellular environment in search of DNA that matches the protospacer, it is the recognition of the PAM sequence that triggers the interference process. Upon identifying the correct PAM, the Cas proteins bind tightly to the target DNA, facilitating subsequent steps in interference. Without the presence of a matching PAM, the Cas proteins will not initiate the cleavage process, ensuring the precision of the immune response.

PAM-Dependent DNA Unwinding

PAM recognition not only initiates interference but also leads to the unwinding of the target DNA. This unwinding is a critical step in preparing the DNA for cleavage. The Cas proteins, once bound to the PAM, induce local distortions in the DNA structure, making it more accessible for further processing. This mechanical aspect of PAM recognition underscores its importance in the CRISPR targeting mechanism.

Variability in PAM Sequences

PAM Sequences in Type II CRISPR-Cas Systems

Type II CRISPR-Cas systems, which include the widely used CRISPR-Cas9, typically have well-defined PAM sequences. For example, the Cas9 protein from Streptococcus pyogenes recognizes a PAM sequence of NGG (where N represents any nucleotide). This simplicity has contributed to the versatility and programmability of Type II CRISPR-Cas systems for genome editing applications.

PAM Sequences in Type I and Other Systems

In contrast, other CRISPR-Cas systems, such as Type I and Type III, exhibit greater variability in their PAM sequences. These systems have evolved to target a broader range of invaders and have more complex PAM recognition patterns. The diversity of PAM sequences in these systems highlights the adaptability of CRISPR-Cas immunity to different environments and challenges.

Engineering PAM Specificity

In the realm of CRISPR-Cas applications, researchers have recognized the significance of PAM sequences. In genome editing, for instance, PAM sequences dictate the range of targetable genomic loci. To expand the utility of CRISPR-Cas systems, scientists have endeavoured to engineer Cas proteins with altered PAM specificities. This innovation has allowed for the targeting of previously inaccessible regions of the genome. Protospacer Adjacent Motifs (PAMs) are the linchpin of CRISPR-Cas target recognition. They ensure the precision of the immune response by distinguishing self from nonself and initiating the interference process. The diversity of PAM sequences across different CRISPR-Cas systems underscores the adaptability and versatility of these prokaryotic immune mechanisms. Harnessing the power of PAM recognition has opened new doors for genome editing and biotechnological applications, further cementing CRISPR-Cas systems as one of the most exciting developments in molecular biology.

5.2 Cascade Complex and Interference Mechanisms

The CRISPR interference stage, a crucial component of the prokaryotic immune system, relies on a multi-protein complex known as Cascade (CRISPR-associated complex for antiviral defence). Cascade plays a central role in target recognition and interference. In this section, we delve into the intricate mechanisms underlying the Cascade complex and how it enables CRISPR-Cas immunity.

Cascade Complex: An Overview

Cascade is a large protein complex composed of multiple subunits, each with distinct functions. It is found in Type I CRISPR-Cas systems, which are among the most widespread in prokaryotes. The Cascade complex acts as a surveillance system, continuously scanning the cell for foreign DNA sequences, or protospacers, that match the sequences stored in the CRISPR array as spacers.

The Composition of Cascade

Cascade typically consists of five to six different proteins, each with its specific role:

Cse1 and Cse2: These proteins form the backbone of the complex. They provide the structural framework for Cascade and are essential for its stability.

Csy1, Csy2, and Csy3 (or Cas5, Cas6, and Cas7): These proteins are responsible for binding the crRNA, or CRISPR RNA. Csy1 interacts directly with the guide crRNA, ensuring its proper positioning within the complex. Csy2 and Csy3 help stabilize the structure.

Cas8: Cas8 proteins, also known as Csc1 in some systems, serve as a platform for binding to target DNA. They play a crucial role in target recognition and interference.

Loading the crRNA

Before Cascade can scan for foreign DNA, it must load the appropriate crRNA, which carries the genetic information of past invaders. This process is highly specific and involves several steps:

crRNA Biogenesis: The primary transcript of the CRISPR array is processed into individual crRNAs, each containing a single spacer sequence. This is usually carried out by a Cas6 endonuclease.

crRNA Binding: Csy1 recognizes and binds to the guide crRNA, ensuring it is oriented correctly within the complex.

Scanning for Matching Protospacers

Once Cascade is loaded with the crRNA, it begins its search for matching protospacers within the cell. This is a critical step in the interference mechanism.

Protospacer PAM Recognition: Cascade scans for a protospacer adjacent motif (PAM) sequence near the target DNA. PAM sequences are short, conserved motifs that are critical for target recognition. They vary between CRISPR-Cas types but are essential for specificity.

Base Pairing: If a PAM sequence is found, Cascade unwinds the DNA duplex and begins base-pairing the crRNA with the target DNA. The crRNA's spacer sequence guides this process, seeking a complementary sequence in the target DNA.

R-Loop Formation: As base-pairing progresses, the crRNA and target DNA form an R-loop structure, with the non-target strand being displaced. This R-loop is a hallmark of successful target recognition.

Interference Mechanism

Having identified a matching protospacer, Cascade initiates the interference process, which ultimately leads to the destruction of the invading DNA.

Recruitment of Cas3: Once the R-loop is formed, Cascade recruits the Cas3 nuclease-helicase complex, a key player in target degradation.

Cas3-Mediated Cleavage: Cas3 possesses nuclease activity, which it uses to cleave the invading DNA. Importantly, it targets the protospacer region, destroying the genetic material of the invader. This step ensures that the foreign DNA is neutralized.

Collateral Cleavage: Cas3 does not discriminate between the target DNA and non-target DNA strands within the R-loop. Consequently, Cascade-mediated interference can lead to collateral cleavage, where both strands of DNA in the vicinity of the protospacer are cleaved, further ensuring the inactivation of the invader.

Feedback Regulation: To prevent self-harm, CRISPR-Cas systems often include mechanisms to avoid targeting their own genetic material. This is achieved through various regulatory mechanisms, such as the exclusion of self-targeting spacers from the CRISPR array or the preferential binding of Cascade to non-self DNA.

Cascade as a Molecular Machine

Cascade is a remarkable molecular machine that carries out multiple functions with precision. Its ability to scan for specific DNA sequences, initiate R-loop formation, and recruit Cas3 for target degradation showcases the sophisticated nature of CRISPR-Cas systems.

Cascade Diversity and Adaptations

It's worth noting that Cascade complexes can vary between different CRISPR-Cas systems, adapting to the specific needs of the host organism. For instance, some variants of Cascade include additional proteins or exhibit structural differences that enhance their target recognition capabilities or provide resistance against anti-CRISPR proteins employed by phages.

Experimental Insights into Cascade Mechanisms

Understanding the Cascade complex and its role in CRISPR interference has been greatly aided by experimental techniques. Cryo-electron microscopy (cryo-EM) and X-ray crystallography have provided high-resolution structures of Cascade, revealing its intricate architecture. These studies have illuminated the positions and functions of individual Cascade proteins, offering critical insights into its mechanisms.

Challenges and Future Directions

While our understanding of Cascade and CRISPR interference mechanisms has grown significantly, many questions remain. Researchers are exploring the variability of Cascade across different CRISPR-Cas systems, the regulation of interference, and the dynamics of target searching and R-loop formation in real-time. These ongoing investigations promise to deepen our knowledge of CRISPR-Cas immunity, potentially unlocking new applications and therapeutic opportunities.

The Cascade complex is a central player in the CRISPR interference mechanism, responsible for target recognition, R-loop formation, and recruitment of the Cas3 nuclease-helicase complex. Its ability to precisely identify and destroy foreign DNA sequences is a testament to the evolutionary ingenuity of CRISPR-Cas systems. As we continue to unravel the complexities of Cascade and its interactions, we move closer to harnessing the full potential of CRISPR technology for various applications in biotechnology and beyond.

5.3 Triggers and Regulation of Interference

In the previous sections, we explored the intricate mechanisms by which prokaryotic CRISPR-Cas systems recognize and target invasive genetic material, leading to interference and the destruction of the invaders. However, for these systems to function effectively and maintain cellular integrity, they must be tightly regulated. In this section, we delve into the fascinating world of how CRISPR interference is triggered and precisely controlled within prokaryotic cells.

Triggering Interference

CRISPR interference is initiated when a prokaryotic cell detects the presence of an invasive genetic element, typically a bacteriophage or plasmid. The recognition of such intruders is a crucial step in the CRISPR-Cas immune response. It involves a two-part process: sensing and interference activation.

Sensing the Invader: One of the primary sensing mechanisms in CRISPR-Cas systems involves the acquisition of spacer sequences during the adaptation stage. These spacer sequences, which are derived from previous encounters with invaders, serve as molecular "mug shots" of potential threats. The cell continuously monitors the genetic material within its environment, comparing it to the stored spacers. If a match is found between a spacer and a segment of the invader's DNA, this acts as a trigger for the CRISPR system to initiate interference.

Protospacer Adjacent Motif (PAM): In addition to spacer matching, many CRISPR-Cas systems require the presence of a specific sequence motif called the Protospacer Adjacent Motif (PAM) for interference to occur. The PAM sequence, which is usually a few nucleotides long and distinct from the spacer, is recognized by the Cas proteins. It serves as a signal that the DNA segment being examined is of foreign origin. Once the PAM is identified, the system can proceed to target and cleave the invasive DNA.

Regulating Interference

While it is crucial for CRISPR-Cas systems to promptly respond to threats, they must also be regulated to prevent unnecessary interference with the host genome. Here, we explore the various levels of regulation that ensure the precise control of CRISPR interference.

Cascade Complex Formation and Activation: A central player in the regulation of CRISPR interference is the Cascade complex, which consists of multiple Cas proteins and crRNAs. This complex not only facilitates target DNA recognition but also acts as a regulatory hub. The activation of Cascade is highly controlled and typically requires the following steps:

crRNA Loading: Before interference can be initiated, the appropriate crRNA must be loaded onto the Cascade complex. This step ensures that only crRNAs matching the detected invader are used for interference.

PAM Recognition: As mentioned earlier, the recognition of a PAM sequence is a critical regulatory checkpoint. Without a PAM match, the CRISPR system cannot proceed with interference.

Cas Protein Activation: Upon successful PAM recognition, the Cascade complex undergoes structural changes that activate the associated Cas proteins. These activated Cas proteins are responsible for cleaving the target DNA.

Small RNAs and Anti-CRISPR Proteins: In addition to the Cascade complex, small RNAs and anti-CRISPR proteins play essential roles in regulating CRISPR interference.

Small RNAs: Some CRISPR systems produce small RNAs from the processed precursor CRISPR RNA (pre-crRNA). These small RNAs can act as regulators, fine-tuning the interference process. They can guide Cas proteins to their targets or modulate the activity of specific components within the system.

Anti-CRISPR Proteins: Intriguingly, phages and other invaders have developed countermeasures to evade CRISPR interference. Anti-CRISPR proteins produced by these invaders can bind to and inhibit key components of the CRISPR-Cas system, effectively shutting down the host's immune response. This interplay between anti-CRISPR proteins and CRISPR-Cas regulation is a dynamic evolutionary arms race.

Timing and Cellular Conditions: CRISPR interference is not a constant process but occurs in response to specific conditions within the cell. Timing and context play significant roles in regulation. For example, interference may be more active during periods of stress or when the cell is actively replicating its DNA. The regulatory mechanisms that sense these conditions and adjust interference accordingly are areas of ongoing research.

Cascade Recycling and Self-Interference Prevention: To prevent self-interference, CRISPR-Cas systems have developed mechanisms to distinguish between the host's own DNA and invasive genetic material. Cascade complexes are often involved in recycling and preventing interference with the host's genome. These recycling mechanisms help maintain genome stability by ensuring that the CRISPR system primarily targets foreign DNA.

Case Study: The Role of Anti-CRISPR Proteins

A fascinating aspect of CRISPR interference regulation is the coevolutionary battle between phages and their prokaryotic hosts. Phages have evolved various strategies to evade CRISPR interference, one of which is the production of anti-CRISPR proteins.

Anti-CRISPR proteins are small proteins encoded by phage genomes. They function by binding to specific Cas proteins or the Cascade complex, inhibiting their activity. These proteins essentially act as molecular decoys, diverting the CRISPR system's attention away from the phage DNA.

Research has shown that anti-CRISPR proteins are remarkably diverse, with different phages producing distinct anti-CRISPRs to counteract the CRISPR-Cas systems of their host bacteria. This diversity reflects the ongoing evolutionary arms race between phages and prokaryotes, where each side continuously develops new tactics to outmanoeuvre the other.

The triggering and regulation of CRISPR interference in prokaryotic cells are highly sophisticated processes. These systems have evolved precise mechanisms to detect invasive genetic material, initiate interference, and prevent self-targeting. The interplay between the CRISPR-Cas system and anti-CRISPR proteins exemplifies the ongoing evolutionary dynamics that shape the battle between prokaryotes and their invaders. Understanding these regulation mechanisms is not only crucial for deciphering the biology of prokaryotes but also holds profound implications for harnessing CRISPR technology in biotechnology and medicine.

Chapter 6: Diversity in CRISPR-Cas Immunity

6.1 Class 1 vs. Class 2 CRISPR-Cas Systems

Classifying CRISPR-Cas systems into different types has been instrumental in understanding the diversity of these prokaryotic immune systems. Two primary classes, Class 1 and Class 2, have emerged as major categories, each with its unique characteristics and mechanisms. In this section, we delve into the distinctions between Class 1 and Class 2 CRISPR-Cas systems, exploring their structural differences, functional implications, and the organisms in which they are commonly found.

Class 1 CRISPR-Cas Systems

Class 1 CRISPR-Cas systems are characterized by their multi-protein complexes and the involvement of multiple effector proteins in the interference stage. Unlike Class 2 systems, which typically rely on a single, large Cas protein such as Cas9 or Cpf1, Class 1 systems employ a more complex machinery, making them intriguing and distinct.

Structure of Class 1 CRISPR-Cas Systems

Class 1 systems are generally larger and more intricate than their Class 2 counterparts. They consist of multiple Cas proteins that form a Cascade (CRISPR-associated complex for antiviral defence) complex. This complex acts as an RNA-guided surveillance complex, patrolling the cell for foreign nucleic acids.

One of the most well-studied Class 1 systems is the Type I CRISPR-Cas system. It is composed of a Cas3 nuclease-helicase and a Cascade complex, which includes several Cas proteins (Cas5, Cas6, Cas7, Cas8, and Cas11). These proteins work together to identify, bind, and degrade invading DNA.

Function of Class 1 CRISPR-Cas Systems

The Class 1 systems have a more versatile and adaptable interference mechanism compared to Class 2 systems. They utilize a crRNA (CRISPR RNA) guide, just like Class 2 systems, but the interference stage involves multiple effector proteins. When a foreign DNA molecule is recognized by the Cascade complex through base-pairing with the crRNA, the Cas3 nuclease-helicase is recruited.

Cas3 then plays a pivotal role in DNA degradation. It cleaves the foreign DNA, ultimately leading to its destruction. This multi-protein approach provides robust immunity against a wide range of invaders and is believed to be particularly effective against mobile genetic elements like plasmids and phages.

Diversity of Class 1 Systems

Class 1 CRISPR-Cas systems are incredibly diverse, with numerous subtypes and variations. Each subtype is associated with different Cas proteins, and these subtypes are often found in various prokaryotic species. For instance, Type I systems are subdivided into several subtypes, including Type I-A, I-B, I-C, and so on, each with its unique set of Cas proteins and characteristics.

This diversity highlights the adaptability of Class 1 systems and their ability to respond to the specific challenges posed by different invaders. It also raises intriguing questions about the evolutionary origins of these systems and how they have diversified over time.

Class 2 CRISPR-Cas Systems

In contrast to the complexity of Class 1 systems, Class 2 CRISPR-Cas systems are characterized by their relative simplicity. These systems rely on a single, large Cas protein as the main effector molecule. One of the most famous examples of a Class 2 system is the Cas9 protein, which has revolutionized genome editing.

Structure of Class 2 CRISPR-Cas Systems

Class 2 systems typically consist of a single Cas protein and a single guide RNA molecule, often referred to as a single-guide RNA (sgRNA). The sgRNA is engineered to target specific DNA sequences through complementary base pairing. In the case of Cas9, this sgRNA directs the protein to the target DNA, where it induces double-strand breaks.

Function of Class 2 CRISPR-Cas Systems

The Class 2 systems, due to their simplicity, offer a straightforward mechanism for DNA interference. Once the Cas protein, guided by the sgRNA, locates the target DNA, it introduces double-strand breaks at precise positions. This break triggers DNA repair mechanisms within the cell, often leading to errors that can result in gene knockout or modification.

Cas9, in particular, has become a powerful tool in genome editing, allowing scientists to precisely manipulate the genomes of a wide range of organisms. This simplicity and versatility have made Class 2 systems highly popular for biotechnological applications.

Diversity of Class 2 Systems

While Class 2 systems are generally simpler than Class 1 systems, they are by no means uniform. Various Class 2 systems have been identified, each with its Cas protein, such as Cpf1 (also known as Cas12) and C2c2 (now known as C2c2/Cas13). These proteins have unique properties, making them valuable for different applications.

For instance, Cpf1 has distinct features that make it well-suited for genome editing in certain organisms, while C2c2/Cas13 has RNA-targeting capabilities, opening up possibilities for RNA editing and RNA interference.

Implications of Class 1 vs. Class 2 Diversity

The coexistence of Class 1 and Class 2 CRISPR-Cas systems in prokaryotes demonstrates the adaptability of these immune systems. Class 1 systems, with their multi-protein complexes, are effective against a wide range of invaders, including plasmids and phages. Their versatility comes at the cost of complexity.

On the other hand, Class 2 systems, with their single Cas proteins and sgRNAs, offer simplicity and precision, making them ideal for genome editing and biotechnological applications. However, they may be less effective against certain invaders, particularly those with mechanisms to evade Cas protein recognition.

Understanding the distinctions between Class 1 and Class 2 systems not only sheds light on the diversity of CRISPR-Cas systems but also informs the development of CRISPR-based technologies. Scientists can choose the most suitable system for a given application based on its characteristics and mechanisms.

Class 1 and Class 2 CRISPR-Cas systems represent two major categories of prokaryotic immune systems, each with its unique structural features and functional mechanisms. The diversity within these classes highlights the adaptability of CRISPR-Cas systems and their ability to provide defence against a wide range of invaders. This knowledge is invaluable for both basic research and the development of CRISPR-based biotechnologies.

6.2 CRISPR Types in Different Prokaryotic Lineages

CRISPR-Cas systems, with their remarkable adaptability and versatility, are not uniform across prokaryotic life. Instead, they exhibit intriguing diversity, with various types and subtypes that have evolved in response to the unique challenges faced by different prokaryotic lineages. In this section, we will delve into the fascinating world of CRISPR diversity, exploring the different types found in various prokaryotic groups and the implications of this diversity.

Classifying CRISPR-Cas Systems

Before we dive into the specifics, it's essential to understand the classification system used to categorize CRISPR-Cas systems. They are broadly categorized into two main classes: Class 1 and Class 2. Each class is further divided into types and subtypes based on the presence of signature Cas proteins and their arrangements.

Class 1 CRISPR-Cas Systems

Class 1 CRISPR-Cas systems are characterized by the involvement of multiple Cas proteins that form a multi-subunit complex to mediate interference. They are typically more extensive and complex than Class 2 systems. Let's explore how Class 1 systems vary among different prokaryotic lineages.

Type I

Type I CRISPR-Cas systems are among the most common Class 1 systems. They are found in a variety of bacteria and archaea, including well-known organisms like *Escherichia coli* and *Mycobacterium tuberculosis*. These systems are known for their distinctive "cascades" of Cas proteins that play a crucial role in target DNA binding and cleavage.

Example 1: E. coli and its Type I CRISPR-Cas System

In *E. coli*, the Type I CRISPR-Cas system involves the Cascade complex, which is composed of multiple Cas proteins and crRNAs. Cascade plays a vital role in the recognition of target DNA sequences, facilitating subsequent interference.

Type III

Type III CRISPR-Cas systems are another Class 1 variety found in various prokaryotes. They have an intriguing dual function, as they can target both RNA and DNA. This dual functionality makes them unique in the CRISPR-Cas world.

Example 2: Type III CRISPR-Cas System in Streptococcus thermophilus

Streptococcus thermophilus, a bacterium used in yogurt production, possesses a Type III-A CRISPR-Cas system. This system not only defends against invasive genetic elements but also plays a role in regulating gene expression, contributing to the bacterium's adaptation to different environments.

Class 2 CRISPR-Cas Systems

Class 2 CRISPR-Cas systems are characterized by a single, large protein (usually Cas9) that takes centre stage in the interference process. They are more streamlined than Class 1 systems and are found in several prominent prokaryotic lineages.

Type II

The Type II CRISPR-Cas system, perhaps the most famous due to its role in genome editing, is associated with the Cas9 protein. It is commonly found in bacteria.

Example 3: Streptococcus pyogenes and Type II CRISPR-Cas

Streptococcus pyogenes, a pathogenic bacterium responsible for strep throat and other infections, harbors a Type II CRISPR-Cas system with the Cas9 protein. Researchers have harnessed this system for precision genome editing applications, revolutionizing the field of molecular biology.

Type V

Type V CRISPR-Cas systems are further divided into several subtypes, with Type V-A and Type V-U being notable representatives. These systems use Cas12 (Cpf1) and C2c2 (Csm) proteins, respectively, for interference.

Example 4: Francisella novicida and Type V-A CRISPR-Cas

Francisella novicida, a highly infectious bacterium, possesses a Type V-A CRISPR-Cas system that employs the Cas12 protein. This system is involved in defence against phage infections and has potential applications in diagnostics due to its collateral cleavage of non-targeted nucleic acids.

Diversity in Archaeal CRISPR-Cas Systems

While much of the CRISPR research has focused on bacterial systems, archaea also harbour unique CRISPR-Cas systems. Archaeal CRISPR systems have distinct features and provide valuable insights into the evolution of these systems.

Example 5: Sulfolobus islandicus and Its Unique CRISPR-Cas System

Sulfolobus islandicus, an extremophilic archaeon, hosts a CRISPR-Cas system that combines features of both Class 1 and Class 2 systems. It showcases the evolutionary flexibility of CRISPR systems, adapting to extreme environmental conditions.

The Impact of CRISPR Diversity

The diversity of CRISPR-Cas systems across prokaryotic lineages has several significant implications:

Adaptive Immunity Strategies

Different prokaryotes face varying selective pressures from viruses and plasmids, leading to the evolution of CRISPR systems tailored to their specific needs. Understanding this diversity provides insights into the adaptive strategies employed by different organisms.

Biotechnological Applications

The discovery of various CRISPR-Cas systems has expanded the toolbox for genome editing and other biotechnological applications. Researchers are exploring the unique properties of different systems for diverse genetic manipulations.

Insights into Evolution

Studying the diversity of CRISPR-Cas systems sheds light on the evolutionary history of prokaryotes and their interactions with mobile genetic elements. This knowledge can help trace the coevolutionary arms race between prokaryotes and their invaders.

CRISPR-Cas systems exhibit remarkable diversity across different prokaryotic lineages, reflecting the rich tapestry of life's interactions with invasive genetic elements. This diversity not only expands our understanding of prokaryotic immunity but also fuels innovations in biotechnology and provides insights into the evolutionary dynamics of these fascinating defence mechanisms.

6.3 Implications for Adaptive Immunity Strategies

In the intricate arms race between prokaryotes and their viral adversaries, the CRISPR-Cas system represents a remarkable breakthrough in the development of adaptive immunity strategies. This subsection explores the implications of CRISPR-Cas systems for the evolution and deployment of adaptive immunity in prokaryotic organisms.

The Versatility of CRISPR-Cas Immunity

One of the key implications of CRISPR-Cas systems for adaptive immunity is their versatility. Unlike the fixed immune systems found in many eukaryotes, such as antibodies or T-cell receptors, CRISPR-Cas systems can adapt rapidly to new threats. This adaptability is a result of the way prokaryotes acquire and incorporate new genetic information into their CRISPR arrays.

CRISPR adaptation, which involves the acquisition of short DNA sequences from invading genetic elements and their integration into the CRISPR array, allows prokaryotes to 'remember' past infections. These acquired sequences, known as spacers, serve as a molecular archive of past encounters with viruses and plasmids. The adaptability of CRISPR-Cas systems is evident in the rapid incorporation of new spacers following an encounter with a novel invader. This ability to update their immune memory gives prokaryotes a significant advantage in the ongoing battle against constantly evolving viral threats.

Specificity and Discrimination

Another profound implication of CRISPR-Cas immunity is its high specificity and discrimination. When prokaryotes integrate a new spacer into their CRISPR array, they also acquire a memory of the invader's genetic material. This allows the CRISPR-Cas system to selectively target and destroy invading nucleic acids bearing sequences matching those of the spacers. The specificity of this process is crucial because it minimizes the risk of mistakenly targeting the host's own genetic material.

The concept of "protospacer adjacent motifs" (PAMs) is central to this discrimination process. PAMs are short, conserved sequences typically found adjacent to the target site in the invader's DNA. The CRISPR-Cas system relies on the presence of a PAM to recognize a valid target. This requirement ensures that the system primarily targets the genetic material of invaders while sparing the host's DNA.

For example, in the type II CRISPR-Cas system, Cas9, a well-known enzyme, uses a PAM sequence as a recognition signal. If the PAM is absent, Cas9 will not bind to the target DNA, preventing nonspecific cleavage. This level of specificity is a key advantage of CRISPR-Cas systems in comparison to other immune mechanisms, which may lack such precise discrimination.

Potential for Genetic Engineering

The precision and versatility of CRISPR-Cas immunity have not gone unnoticed in the field of genetic engineering. Researchers have harnessed the power of CRISPR-Cas systems to develop revolutionary genome editing tools. One of the most famous examples is the CRISPR-Cas9 system, which allows for targeted modifications of genes in a wide range of organisms.

Cas9 is an RNA-guided endonuclease that can be programmed to target specific DNA sequences through the use of synthetic guide RNAs (gRNAs). When the gRNA guides Cas9 to a complementary DNA sequence, Cas9 introduces a double-stranded break at that site. This break activates the cell's repair machinery, which can then be exploited to either disrupt a gene's function or insert new genetic material. This breakthrough technology has opened up a world of possibilities in fields such as medicine, agriculture, and biotechnology.

In medicine, CRISPR-based therapies are being developed to treat genetic diseases. For example, researchers have used CRISPR-Cas9 to correct mutations in the CFTR gene associated with cystic fibrosis. In agriculture, CRISPR is being employed to engineer crops for improved yield, resistance to pests, and enhanced nutritional content. The adaptability of CRISPR-Cas systems means that new applications continue to emerge, demonstrating their immense potential in adaptive immunity strategies.

Coevolutionary Dynamics

Understanding the implications of CRISPR-Cas systems for adaptive immunity also requires an examination of the coevolutionary dynamics between prokaryotes and their viral adversaries. This ongoing arms race drives the constant innovation and adaptation observed in both prokaryotic immune systems and viral evasion strategies.

When prokaryotes develop new spacers against a particular viral sequence, they essentially update their immune arsenal. However, viruses can counter this by undergoing mutations in their genomes, which can render existing spacers ineffective. This prompts prokaryotes to acquire new spacers, leading to a cycle of coevolution. Over time, this dynamic interaction has driven the diversification of CRISPR arrays and the emergence of novel Cas proteins.

A prime example of this coevolution is the discovery of anti-CRISPR proteins produced by certain viruses. These proteins act as inhibitors, blocking the activity of the prokaryotic CRISPR-Cas system and allowing the virus to evade immune detection and destruction. The coevolutionary dance between prokaryotes and their viral adversaries highlights the adaptability of both parties and underscores the ongoing significance of CRISPR-Cas systems in adaptive immunity strategies.

Future Prospects in Adaptive Immunity

As research into CRISPR-Cas systems continues to advance, there is growing anticipation of their potential applications in adaptive immunity strategies beyond prokaryotes. The remarkable adaptability, specificity, and precision of these systems have inspired scientists to explore their use in other organisms, including eukaryotes.

In eukaryotes, where the immune system faces its own challenges in combating diseases like cancer, the prospect of harnessing CRISPR-Cas for adaptive immunity is particularly exciting. Scientists are investigating the development of CRISPR-based therapies for treating genetic disorders and cancer. The ability to selectively target and modify specific genes holds immense promise for personalized medicine.

Additionally, ongoing research seeks to uncover new CRISPR systems in diverse prokaryotic lineages and to understand their unique adaptive immunity strategies. By exploring the full spectrum of CRISPR diversity, scientists aim to gain deeper insights into the coevolutionary dynamics between prokaryotes and their viral adversaries and to discover novel tools for genetic manipulation and disease treatment.

CRISPR-Cas systems have revolutionized our understanding of adaptive immunity in prokaryotes and have opened new avenues for research and applications in diverse fields. Their adaptability, specificity, and potential for genetic engineering make them a powerful tool for combating viral threats and advancing medical and biotechnological frontiers. As research continues to unveil the intricacies of CRISPR-Cas immunity, it is certain that their implications for adaptive immunity strategies will continue to grow and evolve.

Chapter 7: CRISPR Evolution and Horizontal Gene Transfer

7.1 CRISPR Evolutionary Dynamics

The evolution of CRISPR-Cas systems in prokaryotes is a remarkable story of constant adaptation in the face of ever-evolving threats. In this section, we'll explore the dynamic processes that govern the evolution of CRISPR-Cas systems, highlighting key examples and supporting data.

Emergence of New Spacer Sequences

One of the most fascinating aspects of CRISPR evolution is the emergence of new spacer sequences within the CRISPR arrays. Spacers are the genetic memories of past encounters with phages or foreign DNA, and their acquisition is central to the adaptability of CRISPR-Cas systems. Here, we delve into the mechanisms and consequences of spacer acquisition.

Mechanisms of Spacer Acquisition

Spacer acquisition, also known as adaptation, is the process by which prokaryotes capture and integrate short segments of phage or plasmid DNA into their CRISPR arrays. This process is primarily mediated by Cas1 and Cas2 proteins, which form a complex responsible for capturing and integrating the new spacers.

The key steps in spacer acquisition include:

Recognition of foreign DNA: Prokaryotes recognize foreign DNA, such as phage DNA, as a threat.

Capture of protospacers: Cas1 and Cas2 proteins work together to select and capture a short segment of the invading DNA, known as a protospacer.

Integration into CRISPR array: The captured protospacer is integrated into the CRISPR array as a new spacer, positioned between existing spacers.

Evidence of Spacer Acquisition

Spacer acquisition has been extensively studied in various prokaryotes. For instance, a study by Barrangou et al. (2007) investigated the spacer acquisition in Streptococcus thermophilus, a bacterium used in the production of yogurt and cheese. The researchers observed that this bacterium actively acquires new spacers from infecting phages, bolstering its immunity over time. Similar observations have been made in other bacteria.

Moreover, metagenomic studies have provided further evidence of spacer acquisition in natural microbial communities. By sequencing CRISPR arrays from environmental samples, researchers have identified a diverse array of spacers, each representing a historical encounter with a different phage or mobile genetic element.

Coevolutionary Arms Race

The battle between phages and prokaryotes, where phages aim to evade CRISPR immunity and prokaryotes continuously adapt to counteract new threats, resembles an ongoing coevolutionary arms race. This dynamic process has significant implications for the evolution of both CRISPR-Cas systems and phages.

Phage Strategies to Evade CRISPR Immunity

Phages employ various strategies to escape CRISPR-Cas immunity. Some of these strategies include:

Mutation of PAM sequences: Phages can mutate the protospacer adjacent motif (PAM) sequences, making them unrecognizable by the Cas proteins. This renders the phage immune to the prokaryote's CRISPR system.

Anti-CRISPR proteins: Some phages produce anti-CRISPR proteins that inhibit the activity of Cas proteins, allowing them to infect prokaryotic hosts without interference from the CRISPR system.

Escape through recombination: Phages can undergo genetic recombination to generate new sequences that are not targeted by existing CRISPR spacers.

Evidence of Coevolution

The evidence of a coevolutionary arms race between phages and prokaryotes is abundant. Long-term studies of phage-host interactions have shown that as prokaryotes adapt to recognize and target specific phages, the phages, in turn, develop countermeasures to evade the CRISPR system.

Additionally, the discovery of anti-CRISPR proteins provides direct evidence of phage counterstrategies. Bondy-Denomy et al. (2015) identified several anti-CRISPR proteins encoded by phages. These proteins effectively inhibit the CRISPR-Cas system, allowing the phages to establish infection.

CRISPR as a Molecular Fossil Record

One of the most intriguing aspects of CRISPR evolution is that CRISPR arrays serve as a molecular fossil record of past encounters with mobile genetic elements. The spacers within a CRISPR array are like chapters in a book, each telling a story of a previous battle between the prokaryote and an invading genetic element.

Insights from CRISPR Spacer Sequences

Analysing the spacer sequences within CRISPR arrays provides a wealth of information about the historical interactions between prokaryotes and phages. By comparing spacer sequences to known phage genomes and databases, researchers can trace the evolutionary history of phages in specific environments.

Phylogenetic Analysis

Phylogenetic analysis of spacers can reveal patterns of coevolution between prokaryotes and phages. For example, a study by Tyson et al. (2004) examined the spacers in the

CRISPR arrays of archaea from acidic hot springs. The researchers found that the spacers matched phage sequences from the same environment, indicating an ongoing arms race.

Implications for Microbial Communities

CRISPR-Cas systems are not limited to individual prokaryotes; they also play a significant role in shaping microbial communities. The evolution of CRISPR immunity within a community can influence the composition and dynamics of that community.

Community-Level Adaptation

In microbial communities, the acquisition of new spacers by one member can benefit the entire community. For example, if one bacterium acquires a spacer that targets a common phage, it provides protection not only for itself but also for neighbouring bacteria through the release of phage fragments.

Diversity and Competition

CRISPR-mediated immunity can lead to the diversification of microbial communities. Bacteria with different spacer repertoires may coexist because they are immune to different phages, reducing competition for resources.

The evolutionary dynamics of CRISPR-Cas systems in prokaryotes are a testament to the ingenuity of nature's defence mechanisms. Through the ongoing acquisition of new spacers, the coevolutionary arms race with phages, and the preservation of molecular fossils in the form of spacer sequences, CRISPR-Cas systems continue to adapt and thrive in the face of evolving threats. These mechanisms not only safeguard individual prokaryotes but also have profound implications for microbial communities and ecosystems. Further research in this field promises to unveil even more insights into the fascinating world of CRISPR-Cas evolution.

7.2 How CRISPR Immunity Influences Horizontal Gene Transfer

Horizontal Gene Transfer (HGT) is a fundamental mechanism that allows prokaryotic organisms to acquire new genetic material from their environment, leading to rapid adaptation and evolution. CRISPR-Cas systems, originally evolved as a prokaryotic defence mechanism against invading mobile genetic elements like plasmids and phages, play a pivotal role in shaping the dynamics of HGT within microbial communities. In this subsection, we will explore how CRISPR immunity influences HGT, using examples and data to illustrate these complex interactions.

CRISPR as a Barrier to Horizontal Gene Transfer

CRISPR-Cas systems are often regarded as a barrier to HGT because they can specifically target and degrade incoming foreign DNA, such as plasmids or phages. When a prokaryote acquires a new spacer sequence that matches a region of the invading genetic material, the CRISPR-Cas system is activated, leading to the cleavage and degradation of the invader DNA. This process can significantly hinder the successful acquisition of new genetic material.

Example 1: Plasmid Resistance

A classic example of CRISPR-mediated interference in HGT is seen in Escherichia coli. When this bacterium possesses a CRISPR-Cas system that targets a specific plasmid, it can resist plasmid invasion. Studies have shown that the presence of an active CRISPR system in E. coli can lead to a significant reduction in the successful transfer of plasmids harbouring targeted sequences.

CRISPR-Enhanced Immunity

Notably, the presence of CRISPR-Cas systems can go beyond merely acting as a barrier. It can enhance the immunity of prokaryotes against specific genetic elements. This is particularly evident when CRISPR spacers match essential regions of mobile genetic elements, rendering these elements less effective.

Example 2:Phage Immunity

In a study of Streptococcus thermophilus, researchers found that strains with functional CRISPR-Cas systems exhibited higher resistance to phage infection compared to strains lacking CRISPR. The specific spacers in the CRISPR array targeting the phage's critical genes were found to be highly effective in preventing phage proliferation, providing a significant advantage to the host strain.

CRISPR-Driven Evolution of Mobile Genetic Elements

While CRISPR systems can act as barriers and enhance immunity, they also drive the evolution of mobile genetic elements, such as phages and plasmids. This occurs through a continuous coevolutionary arms race between prokaryotes and their mobile genetic element adversaries.

Example 3: Coevolution of Phages

A study tracking the interaction between Pseudomonas aeruginosa and its phage adversaries demonstrated that the phages underwent rapid genetic mutations to evade the CRISPR-Cas system. Over time, phages with mutations in their protospacer regions or the presence of anti-CRISPR proteins became more prevalent in the population. This illustrates how CRISPR-Cas systems drive the diversification of mobile genetic elements.

HGT Promotion by CRISPR-Tolerant Elements

Interestingly, CRISPR-Cas systems have also been found to promote HGT under certain conditions. When invading genetic elements carry anti-CRISPR proteins or other evasion mechanisms, they can escape CRISPR-mediated immunity. This phenomenon can enable the successful transfer of genetic material.

Example 4: Anti-CRISPR Proteins

Anti-CRISPR proteins produced by some phages have the ability to inhibit the activity of the host CRISPR-Cas system temporarily. Consequently, these phages can evade the immune response, facilitating their integration into the host genome. In a study of Pseudomonas aeruginosa, it was discovered that certain phages carried genes encoding anti-CRISPR proteins, allowing them to overcome the host's CRISPR immunity and promoting successful HGT.

Balancing Selection in CRISPR-Cas Systems

The complex interplay between CRISPR-Cas systems and mobile genetic elements results in balancing selection. This dynamic equilibrium maintains genetic diversity within microbial populations.

Example 5: Diversity in Spacer Sequences

A comprehensive analysis of spacer sequences in microbial communities reveals a fascinating diversity of spacers targeting various genetic elements. Some spacers are highly conserved and effective against common invaders, while others are rapidly evolving due to the arms race between CRISPR and mobile genetic elements. This diversity highlights the ongoing adaptation and selection pressures on CRISPR-Cas systems.

Ecological Implications of CRISPR-Modulated HGT

The influence of CRISPR immunity on HGT has significant ecological implications. It affects not only the evolutionary trajectories of individual prokaryotic species but also the structure and functioning of entire microbial communities.

Example 6: Community-Level Effects

In a study of a microbial consortium involved in wastewater treatment, researchers observed that the presence of CRISPR-Cas systems influenced the composition of the microbial community. Species with active CRISPR-Cas systems exhibited lower rates of plasmid transfer, leading to differences in community structure and function compared to communities lacking CRISPR-based immunity.

Implications for Biotechnology and Medicine

Understanding how CRISPR immunity influences HGT has practical implications in biotechnology and medicine, particularly in the development of strategies to control antibiotic resistance and manipulate microbial communities.

Example 7: Combating Antibiotic Resistance

The ability of CRISPR-Cas systems to hinder the spread of antibiotic resistance genes through plasmids has led to investigations into using CRISPR-based strategies to combat antibiotic-resistant bacteria. By designing CRISPR spacers to target specific resistance genes, researchers aim to prevent their transfer, potentially providing a novel approach to addressing the global antibiotic resistance crisis.

CRISPR-Cas systems play a multifaceted role in shaping the dynamics of horizontal gene transfer within prokaryotic communities. While they act as barriers against invading genetic elements, they also drive the evolution of mobile genetic elements, leading to a continuous coevolutionary arms race. Understanding these intricate interactions is crucial not only for advancing our knowledge of prokaryotic immunity but also for harnessing CRISPR-based strategies for biotechnological and medical applications. The examples and data presented here underscore the complexity and significance of CRISPR-mediated regulation of HGT in microbial ecosystems.

7.3 Coevolution of Phages and Prokaryotes

The intricate dance of coevolution between bacteriophages (phages) and prokaryotes has been a driving force shaping the evolution of CRISPR-Cas immune systems in prokaryotes. This coevolutionary arms race has not only profoundly impacted the genetic diversity of both phages and their bacterial hosts but has also provided a fascinating window into the dynamics of molecular warfare at the microscopic level.

Understanding Phage-Host Coevolution

Coevolution is a dynamic process where two or more species exert selective pressures on each other, leading to reciprocal adaptations over time. In the context of phages and prokaryotes, this coevolutionary interaction is characterized by a continuous struggle for dominance and survival. Here, we delve into the mechanisms and consequences of this intriguing phenomenon:

Phage Adaptations to Bypass CRISPR Defence

Mutational Escapes: Phages have evolved mechanisms to evade CRISPR defences. One common strategy is mutational escape, where phages accumulate mutations in the protospacer region targeted by CRISPR spacers. This makes them less recognizable by the Cas machinery. An example is the Pseudomonas aeruginosa phage DMS3, which rapidly mutates its DNA to avoid detection by the host's CRISPR-Cas system.

Anti-CRISPR Proteins: Some phages produce anti-CRISPR proteins. These proteins can inhibit the Cas enzymes, preventing them from interfering with the phage's replication. A well-known example is AcrF1, an anti-CRISPR protein produced by phage phiKZ, which inhibits the type I-F CRISPR-Cas system in Pseudomonas aeruginosa.

Prokaryotic Counter-Adaptations

Spacer Acquisition: In response to phage evasion tactics, prokaryotes engage in counter-adaptations. They acquire new spacers that specifically target the evolved phage strains. For instance, studies have shown that when Escherichia coli was challenged with a phage carrying a single-point mutation in the protospacer, the bacteria quickly adapted by acquiring new spacers that matched the mutated sequence.

Diverse CRISPR Types: The diversity of CRISPR-Cas systems across different prokaryotes is thought to be partially driven by phage-host coevolution. Some CRISPR types may be more effective against certain phages, prompting hosts to diversify their immune strategies.

A Red Queen's Race

The dynamics of phage-host coevolution resemble the "Red Queen's Race" from Lewis Carroll's Through the Looking-Glass, where characters must run as fast as they can just to stay in the same place. In this biological arms race, phages continuously evolve to infect their hosts, while prokaryotes, in turn, evolve to resist these evolving phages. This rapid adaptation on both sides contributes to the remarkable genetic diversity observed in natural populations of both phages and prokaryotes.

The Genetic Footprint of Coevolution

The coevolutionary arms race between phages and prokaryotes leaves distinct genetic signatures that scientists have uncovered through extensive research. These signatures shed light on the mechanisms and consequences of this coevolutionary process:

Coevolutionary Hotspots

CRISPR Spacer Diversity: In prokaryotic genomes, the diversity of CRISPR spacers is often concentrated in regions associated with phage-host interactions. This clustering of spacers indicates active selection for immunity against prevalent phages.

Phage Genome Diversity: Phage genomes also show evidence of coevolution. Phages frequently carry sequences resembling known protospacers, suggesting an ongoing battle with their host's CRISPR-Cas system.

Genome Plasticity

Bacterial Mutations: The need to constantly adapt to phage attacks has driven prokaryotes to evolve rapidly. Mutations in key components of CRISPR-Cas systems, such as Cas proteins or regulatory elements, are often observed. These mutations can enhance immunity or alter the specificity of the system.

Phage Host Range: Phages themselves exhibit genome plasticity, with frequent mutations and horizontal gene transfer. This enables them to infect a broader range of hosts and evade immune systems.

Ancient CRISPR Arrays

Palaeobiological Insights: The study of ancient DNA has provided a glimpse into the deep history of phage-host interactions. Researchers have identified ancient CRISPR arrays in ancient prokaryotic genomes, hinting at the longstanding coevolutionary battle.

Implications for Ecology

Microbial Ecosystem Dynamics: The coevolution of phages and prokaryotes has far-reaching consequences for microbial ecosystems. These dynamics influence the composition and diversity of microbial communities, which, in turn, impact nutrient cycling, disease dynamics, and biogeochemical processes.

The Future of Coevolutionary Research

As our understanding of the coevolution between phages and prokaryotes continues to deepen, it raises exciting questions about the potential applications and implications of this knowledge:

Biotechnological Applications

Phage Therapy: Can we leverage our understanding of phage-host coevolution to improve phage therapy, using phages that have coevolved with specific bacterial strains to enhance treatment efficacy?

Antibiotic Alternatives: How might coevolutionary insights inform the development of alternative antimicrobial strategies, especially in the face of antibiotic resistance?

Ecological Dynamics

Microbial Community Stability: How do coevolutionary dynamics influence the stability and resilience of microbial communities, and what are the broader ecological consequences?

Evolutionary Principles

Generalizable Lessons: Can we extract general principles of coevolution from the study of phages and prokaryotes that apply to other host-pathogen systems?

The coevolution of phages and prokaryotes is a captivating tale of molecular warfare that has left an indelible mark on the genetic landscapes of both parties. Understanding the mechanisms and consequences of this coevolutionary arms race not only sheds light on the intricate dance of life at the microscopic level but also holds promise for biotechnological applications and ecological insights that extend far beyond the world of bacteria and their viral adversaries. The ongoing exploration of this dynamic relationship promises to uncover new surprises and deepen our appreciation of the complex web of life.

Chapter 8: CRISPR-Cas Applications and Biotechnology

8.1 The CRISPR Revolution in Genome Editing

In recent years, the field of genetics and molecular biology has witnessed a seismic shift, thanks to the advent of CRISPR-Cas (Clustered Regularly Interspaced Short Palindromic Repeats and CRISPR-associated proteins) genome editing technology. This revolutionary system has unlocked unprecedented precision and efficiency in manipulating the genetic code of organisms. In this section, we will explore the profound impact of CRISPR genome editing, examining its mechanisms, applications, and the remarkable strides it has made in diverse fields.

Understanding the Mechanism: How CRISPR Genome Editing Works

Before diving into the far-reaching applications, it's essential to grasp the basic mechanism behind CRISPR genome editing. At its core, the CRISPR-Cas system serves as a defence mechanism in bacteria and archaea against viral invaders. It functions by using RNA molecules to guide Cas proteins to precise locations in the genome, where they can make targeted cuts in the DNA. This mechanism, harnessed for genome editing, is often referred to as the "molecular scissors" of genetics.

The key components of the CRISPR-Cas genome editing system include:

Guide RNA (gRNA): A synthetic RNA molecule designed to match the DNA sequence to be edited.

Cas9 Protein: The endonuclease responsible for making the DNA cuts.

Repair Mechanisms: The cell's natural repair mechanisms, such as Non-Homologous End Joining (NHEJ) or Homology-Directed Repair (HDR), are employed to introduce changes at the cut site.

When the gRNA and Cas9 protein are introduced into a cell, they form a complex that navigates to the specified DNA sequence, guided by the complementary RNA. Once the target DNA is located, Cas9 introduces a precise double-strand break. The cell then attempts to repair this break, often introducing small insertions or deletions (indels) in the process. By controlling this repair process, scientists can induce gene disruptions or insert new genetic material, enabling precise genome modifications.

Applications in Medicine: Treating Genetic Diseases

One of the most celebrated achievements of CRISPR genome editing is its potential to treat genetic diseases. It offers the tantalizing prospect of correcting the genetic mutations responsible for a wide range of inherited disorders. For example, in 2019, scientists successfully used CRISPR-Cas to treat sickle cell anaemia in mice. This work involved modifying the mutated haemoglobin gene, effectively curing the disease in these animals. While clinical applications in humans are still in early stages, the implications for patients with genetic disorders are profound.

Agriculture: Enhancing Crop Resilience and Yield

Beyond medicine, CRISPR genome editing has transformative implications for agriculture. Crop scientists are using this technology to develop new strains of crops that are more resilient to environmental stresses, require fewer pesticides, and yield higher-quality produce. A striking example is the development of a drought-resistant strain of rice, which holds great promise in addressing food security challenges in regions prone to water scarcity.

Biotechnology: Advancements in Bioprocessing

The biotechnology industry has embraced CRISPR genome editing for various applications. Researchers are modifying microorganisms to produce valuable compounds such as biofuels, pharmaceuticals, and industrial chemicals with greater efficiency. This technology allows for precise control over metabolic pathways, enhancing the production of desired products while reducing waste.

Conservation: Protecting Endangered Species

CRISPR genome editing also extends its reach to conservation efforts. Scientists are exploring the possibility of using this technology to rescue endangered species by correcting genetic defects that threaten their survival. For instance, the gene-editing of coral species to make them more resilient to climate change impacts could play a pivotal role in preserving coral reefs, which are vital ecosystems for marine life.

Ethical Considerations and Regulatory Oversight

With the power to edit genes comes significant ethical responsibility. The ability to modify the human germline, for instance, raises profound ethical questions about the potential for designer babies and unintended consequences. To address these concerns, there is a growing need for clear and robust ethical guidelines and regulatory oversight. Many countries have established regulations governing the use of CRISPR technology, particularly in the context of human genome editing.

Challenges and Limitations

While the CRISPR genome editing revolution is undeniably groundbreaking, it is not without its challenges and limitations. One key challenge is off-target effects, where Cas9 may unintentionally modify DNA sequences similar to the target. Researchers are continually working to enhance the precision of the technology to minimize such errors.

Another limitation is the size of DNA sequences that can be edited. Cas9 has size constraints, making it challenging to edit longer stretches of DNA. However, ongoing research is exploring alternative CRISPR systems with larger editing capacities.

Future Prospects and Emerging Trends

As the field of CRISPR genome editing continues to evolve, several exciting trends and prospects are on the horizon:

Prime Editing: A novel technique that allows for precise and highly efficient editing with minimal off-target effects, showing great promise for therapeutic applications.

Epigenome Editing: The ability to modify epigenetic marks on DNA, which can influence gene expression without altering the underlying DNA sequence.

Multiplex Editing: Advancements enabling simultaneous editing of multiple genes, opening up new possibilities for complex genetic modifications.

In Vivo Therapies: Developing in vivo delivery methods to treat diseases directly within the human body, reducing the need for ex vivo editing.

The CRISPR genome editing revolution has transformed the way we think about genetics, offering unprecedented opportunities to modify DNA with precision and efficiency. Its applications span medicine, agriculture, biotechnology, conservation, and beyond. However, the ethical considerations and technical challenges associated with this powerful tool demand careful consideration. As CRISPR technology continues to advance, its impact on science and society will undoubtedly shape the future of genetic research and biotechnology.

8.2 Beyond Genome Editing: Other Biotechnological Applications

While the development of CRISPR-Cas for genome editing has been groundbreaking, this versatile technology extends far beyond the realm of genetic manipulation. Researchers and biotechnologists have harnessed the power of CRISPR-Cas systems for a wide array of applications, revolutionizing fields such as diagnostics, biotechnology, and agriculture. In this subsection, we explore some of the most exciting and impactful non-genome editing applications of CRISPR-Cas systems, backed by examples and data that highlight their potential.

CRISPR-Based Diagnostic Tools

One of the most promising areas of non-genome editing CRISPR applications lies in diagnostics. The precision and specificity of CRISPR-Cas systems make them invaluable for detecting nucleic acids, including DNA and RNA from pathogens. Here are some key examples:

Detection of Infectious Diseases

CRISPR-based diagnostic tools, such as the SHERLOCK (Specific High Sensitivity Enzymatic Reporter UnLOCKing) system, have demonstrated exceptional sensitivity and speed in detecting infectious agents. For instance, researchers have successfully used SHERLOCK to identify Zika and Dengue viruses in patient samples with detection limits as low as a single copy of viral RNA per microliter. The ability to quickly and accurately diagnose infectious diseases is crucial for timely treatment and containment.

Cancer Biomarker Detection

In oncology, CRISPR-Cas technology has been employed to develop highly sensitive and specific assays for detecting cancer-related mutations and biomarkers. Examples include the CRISPR/Cas12-based DETECTR (DNA Endonuclease-Targeted CRISPR Trans Reporter) and HOLMES (one-HOur Low-cost Multipurpose highly Efficient System) systems. These assays can detect cancer-related mutations with high accuracy, opening doors to early cancer detection.

Precision Agriculture

The agriculture industry has been quick to embrace CRISPR-Cas technology for various applications, ranging from crop improvement to pest management:

Crop Improvement

CRISPR-Cas systems are employed to create genetically modified (GM) crops with enhanced traits, such as disease resistance, improved yield, and nutritional content. A remarkable example is the development of non-browning mushrooms. Researchers used CRISPR to edit the genome of white button mushrooms, reducing browning and increasing shelf life. These mushrooms could help reduce food waste and enhance the quality of agricultural produce.

Pest Control

In the battle against agricultural pests, CRISPR-Cas technology offers a precise and environmentally friendly solution. Scientists are exploring the use of CRISPR to genetically modify insects like mosquitoes to reduce their ability to transmit diseases. For example, researchers have engineered mosquitoes with reduced susceptibility to malaria parasites using CRISPR-Cas9. Such innovations hold great promise for disease control in regions affected by mosquito-borne illnesses.

Antibiotic Resistance Mitigation

The global challenge of antibiotic resistance has prompted creative solutions, and CRISPR-Cas systems offer an innovative approach:

Phage Therapy

Phage therapy, the use of bacteriophages to treat bacterial infections, has gained renewed interest due to the rise of antibiotic-resistant bacteria. CRISPR-Cas systems have been harnessed to precisely target and eliminate antibiotic resistance genes in bacterial genomes. In a study published in Nature Communications, researchers successfully used

CRISPR-Cas9 to combat antibiotic-resistant Staphylococcus aureus in mice, demonstrating the potential of this approach for treating infections.

Environmental Applications

CRISPR-Cas systems have also found applications in environmental science and conservation:

Conservation Genetics

Conservationists are utilizing CRISPR technology to aid endangered species' preservation efforts. For example, researchers are exploring the use of CRISPR-Cas to modify the genomes of coral species threatened by climate change. By enhancing their thermal tolerance, coral reefs may have a better chance of surviving rising ocean temperatures.

Environmental Monitoring

CRISPR-Cas systems have been employed to develop biosensors for monitoring environmental pollutants. These biosensors can detect specific contaminants in water or air, offering rapid and targeted detection methods. This technology is essential for environmental protection and ensuring the safety of ecosystems.

Therapeutic Applications Beyond Gene Editing

Beyond diagnostics, CRISPR-Cas systems have the potential to revolutionize therapeutic approaches in various ways:

Targeted Drug Delivery

Researchers are exploring CRISPR-Cas technology to develop targeted drug delivery systems. By using CRISPR-Cas to modify immune cells or other cell types, drugs can be delivered precisely to disease sites, reducing side effects and improving treatment efficacy. This approach holds promise for cancer therapy and autoimmune disease treatment.

Disease Modelling

CRISPR-Cas systems are instrumental in creating disease models for research and drug development. Organoids and animal models with precisely engineered disease-related mutations enable researchers to better understand disease mechanisms and test potential therapies. For example, CRISPR-edited organoids have been used to study diseases like cystic fibrosis.

The applications of CRISPR-Cas systems extend far beyond genome editing, encompassing diagnostics, agriculture, antibiotic resistance mitigation, environmental science, and therapeutic innovations. The examples and data presented here underscore the transformative potential of CRISPR technology in addressing some of the most pressing challenges in healthcare, agriculture, and environmental conservation. As research in this field continues to advance, we can anticipate even more groundbreaking applications that harness the precision and versatility of CRISPR-Cas systems for the benefit of humanity and the environment.

8.3 Ethical and Regulatory Considerations

In the realm of biotechnology, few innovations have sparked as much excitement and controversy as the CRISPR-Cas technology. While its potential to revolutionize medicine, agriculture, and various other fields is undeniable, it also raises profound ethical and regulatory questions. This subsection delves into the complex landscape of ethical and regulatory considerations surrounding CRISPR-Cas applications.

Off-Target Effects and Genome Integrity

One of the foremost ethical concerns associated with CRISPR-Cas technology is the potential for off-target effects, where the Cas proteins may inadvertently edit genes other than the intended target. The accuracy of CRISPR systems has improved significantly, but it's not foolproof. Studies have identified off-target effects in various organisms, raising questions about the long-term consequences of unintended genetic changes.

For instance, a study published in **Nature Biotechnology** in 2020 demonstrated that in certain cases, CRISPR-Cas9 can induce large deletions or rearrangements at off-target sites. These unintended changes could lead to unexpected health consequences in gene therapy or genetically modified organisms (GMOs). As CRISPR applications move from the laboratory to the clinic and fields, ensuring the precision and safety of the technology becomes paramount.

Germline Editing and Hereditary Changes

Perhaps one of the most ethically charged debates surrounding CRISPR-Cas revolves around the editing of human germline cells. This process, known as germline editing, has the potential to eliminate hereditary genetic diseases from future generations. However, it also raises concerns about the permanence and unforeseen consequences of such genetic alterations.

The famous case of the "CRISPR babies" in China in 2018 exemplifies the ethical minefield of germline editing. The Chinese scientist He Jiankui claimed to have edited the genes of twin girls to make them resistant to HIV. The international scientific community widely condemned this action as reckless and premature due to the unknown long-term effects of the genetic modifications and a lack of ethical oversight.

Access and Equity

CRISPR-Cas technology has the potential to exacerbate global inequalities in access to healthcare and biotechnology. The high costs associated with gene therapies and genetic modifications could create disparities between those who can afford these treatments and those who cannot. This raises concerns about social justice and equitable access to the benefits of CRISPR-Cas advancements.

For example, the gene-editing therapy Zolgensma, which uses CRISPR-related technology, was approved in the United States for the treatment of spinal muscular atrophy (SMA). However, the treatment comes with an astronomical price tag, limiting access for many SMA patients around the world. Ethical discussions often centre on the need for policies that ensure the fair distribution of CRISPR-based therapies.

Dual-Use Dilemma

CRISPR technology has dual-use potential, meaning it can be used for both beneficial and harmful purposes. This dual-use dilemma poses a significant ethical challenge, particularly in the context of biosecurity and bioterrorism. While CRISPR has the potential to address public health challenges, such as creating disease-resistant crops, it also has the potential to create bioweapons, such as drug-resistant pathogens.

The ease of obtaining CRISPR technology components and the democratization of gene editing tools raise concerns about the responsible use of this technology. International agreements and regulations must be developed and enforced to prevent malicious applications.

Regulatory Frameworks

The regulation of CRISPR-Cas technology varies widely from one country to another. While some nations have implemented strict guidelines and regulatory frameworks, others have adopted more permissive approaches. This patchwork of regulations creates challenges in overseeing and controlling the technology's use.

In the United States, for instance, the FDA (Food and Drug Administration) regulates CRISPR-based therapies, but the regulatory process can be complex and lengthy. Meanwhile, countries like China have moved more quickly to embrace CRISPR applications, which has led to concerns about oversight and safety.

International bodies like the World Health Organization (WHO) have initiated efforts to develop global standards for the use of gene editing technologies, including CRISPR-Cas. The aim is to establish a unified approach that addresses ethical concerns while fostering scientific progress.

Public Engagement and Informed Consent

Public engagement and informed consent are crucial components of responsible CRISPR-Cas research and applications. As decisions are made about the use of CRISPR technology in various contexts, it's essential to include diverse stakeholders in the decision-making process.

For instance, when conducting clinical trials involving CRISPR-Cas gene therapies, ensuring that participants fully understand the potential risks and benefits is essential. In agricultural applications, farmers and consumers should have a say in the use of genetically modified crops.

Ethical Oversight and Responsible Innovation

The ethical and regulatory considerations surrounding CRISPR-Cas technology are multifaceted and continuously evolving. As the technology advances, so too must the ethical frameworks and regulatory mechanisms that govern its use.

Responsible innovation, guided by ethical principles, will be crucial in harnessing the potential of CRISPR-Cas for the benefit of humanity while mitigating risks. It is a collective responsibility of scientists, policymakers, ethicists, and the public to ensure that CRISPR-Cas applications are used in ways that align with societal values, promote equity, and safeguard the integrity of the genome.

In this ever-evolving landscape, the development of thoughtful, adaptable, and internationally coordinated ethical and regulatory frameworks is essential to navigate the promising yet challenging path paved by CRISPR-Cas technology.

Chapter 9: CRISPR in Microbial Ecology

9.1 CRISPR as a Tool for Microbial Community Analysis

In the vast and complex world of microbial ecology, understanding the composition and dynamics of microbial communities is essential. Microbes, including bacteria, archaea, viruses, and fungi, play critical roles in ecosystems, ranging from soil and water to the human gut. Their activities impact nutrient cycling, disease dynamics, and even climate change. To decipher the intricacies of these microbial communities, scientists have turned to an unexpected ally: CRISPR-Cas systems.

The Complexity of Microbial Communities

Microbial communities are incredibly diverse, with numerous species coexisting and interacting within a given environment.

Traditional techniques, such as DNA sequencing and culturing, have been valuable for identifying individual microbial species. However, they often fall short in capturing the intricate web of interactions that govern microbial ecosystems.

Understanding these complex communities requires technologies that can provide insights into the identities of microorganisms, their functions, and the intricate networks of ecological interactions. CRISPR-based approaches have emerged as powerful tools for unravelling the mysteries of microbial communities.

CRISPR and Microbial Community Analysis: A Synergistic Approach

One of the most innovative applications of CRISPR in microbial ecology is its use in metagenomics. Metagenomics involves the study of genetic material recovered directly from environmental samples, without the need for cultivation. This approach allows researchers to examine the genetic material of entire microbial communities, providing a holistic view of their composition and functional potential.

Here's how CRISPR is harnessed in microbial community analysis:

Profiling Microbial Diversity with CRISPR-Spacer Sequencing

CRISPR arrays in bacterial and archaeal genomes serve as a molecular memory of past encounters with invading genetic elements, such as viruses and plasmids. Each CRISPR array contains short, unique sequences called spacers, which are derived from these invaders. By sequencing and analysing the spacers in a microbial community, researchers can gain insights into the historical interactions between microorganisms and their viral predators.

For example, a study conducted in a hypersaline lake ecosystem revealed the diversity of viruses present by sequencing the spacers in the CRISPR arrays of the resident archaea. This approach not only identified known viruses but also discovered previously unknown viral species, shedding light on the arms race between microbes and their viral foes.

Reconstruction of Microbial Genomes

CRISPR sequences provide a unique opportunity to "target" specific microorganisms in a community. Researchers can design CRISPR guides to match the sequences of interest and use Cas proteins to cleave the DNA of the targeted microbe. This approach, known as CRISPR-based microbial genome reconstruction, enables the assembly of genomes from individual microorganisms within a complex community.

A notable example comes from the study of the human gut microbiome. By using CRISPR to selectively enrich for specific bacterial genomes, researchers have been able to obtain high-quality genomic data from elusive and previously unculturable gut bacteria. This has greatly expanded our understanding of the functions and roles of these microbes in human health.

Tracking Microbial Interactions

CRISPR spacers not only reveal historical interactions but can also be used to probe ongoing microbial relationships. Researchers can design synthetic CRISPR systems that target specific microorganisms or genes within a community. By monitoring changes in the abundance of specific spacers, they can infer ecological interactions, such as predation or competition, between microorganisms.

For instance, a study in soil ecosystems used CRISPR-based tracking to elucidate the predation dynamics of predatory bacteria on other microbial species. This revealed intricate networks of predation, shedding light on the factors shaping soil microbial communities.

Functional Analysis of Microbial Communities

Beyond identifying species and interactions, CRISPR can be employed to understand the functional potential of microbial communities. Researchers can design CRISPR-based assays to target specific functional genes, such as those involved in nitrogen fixation or methane metabolism, and assess their abundance and activity within a community.

In a study of a nitrogen-fixing microbial community in the ocean, CRISPR-based assays were used to measure the expression of key nitrogenase genes. This provided insights into the factors controlling nitrogen fixation rates in marine ecosystems and their implications for global nutrient cycles.

Challenges and Future Directions

While CRISPR has opened new frontiers in microbial community analysis, challenges remain. These include the need for more robust bioinformatics tools to analyse complex metagenomic datasets, as well as ethical considerations when manipulating microbial communities using CRISPR technologies.

The future of CRISPR in microbial ecology holds promise. As sequencing technologies continue to advance and our understanding of CRISPR systems deepens, researchers are likely to uncover even more sophisticated applications for deciphering the mysteries of microbial communities. Additionally, CRISPR-based approaches may help us develop targeted interventions to mitigate the impacts of harmful microbes or enhance the functions of beneficial ones, with implications for fields ranging from agriculture to environmental conservation.

CRISPR-Cas systems have revolutionized microbial community analysis by providing a versatile toolkit for profiling diversity, reconstructing genomes, tracking interactions, and assessing functional potential. These applications have transformed our ability to explore the hidden world of microbes, shedding light on their roles in ecosystems and their relevance to human health and environmental sustainability.

9.2 Impacts of CRISPR Immunity on Microbial Ecosystems

Microbial ecosystems, whether they inhabit the ocean depths, the human gut, or soil ecosystems, are incredibly diverse and complex. Within these ecosystems, the interactions between microorganisms play a crucial role in shaping their composition and functioning. In recent years, it has become increasingly clear that the presence and activity of CRISPR-Cas systems in various microorganisms have significant impacts on these ecosystems. This section explores the multifaceted impacts of CRISPR immunity on microbial ecosystems, using examples and data to illustrate these effects.

Controlling Bacterial Abundance

One of the most direct and well-documented impacts of CRISPR immunity in microbial ecosystems is its role in controlling the abundance of specific bacterial species, particularly those targeted by CRISPR-Cas systems. Let's consider an example in a marine ecosystem.

Example 1: Marine Ecosystems

In the marine environment, viruses known as bacteriophages infect and replicate within bacterial hosts, affecting bacterial populations. CRISPR-Cas systems provide a defence mechanism against these phages. A study conducted in the Baltic Sea revealed that the presence of CRISPR-Cas systems in marine bacteria significantly reduced the abundance of phages. Researchers observed that bacterial populations equipped with functional CRISPR-Cas systems had a competitive advantage over those lacking these systems. This advantage resulted in a lower incidence of viral infections, which, in turn, influenced the overall structure of microbial communities in the Baltic Sea. These findings highlight how CRISPR immunity can directly impact the abundance and distribution of bacterial species in marine ecosystems.

Biodiversity and Coexistence

CRISPR immunity can also have indirect effects on microbial biodiversity within ecosystems. By reducing the abundance of specific bacterial strains, CRISPR-Cas systems can create niches for other microbial species to thrive. This phenomenon is particularly evident in environments where resource competition is intense.

Example 2: Soil Microbial Communities

In soil ecosystems, a vast array of microorganisms, including bacteria, archaea, fungi, and viruses, coexist and interact. Soil bacteria equipped with CRISPR-Cas systems can selectively target and inhibit the growth of closely related bacterial strains, reducing competition for resources. This selective pressure can lead to the coexistence of diverse bacterial populations in soil, each with its own set of CRISPR-Cas systems targeting different phages and competitors. The net effect is an increase in microbial diversity, as observed in studies analysing soil metagenomes.

Ecological Resilience

CRISPR-Cas systems can also contribute to the ecological resilience of microbial ecosystems by preventing the dominance of a single species. This diversity is crucial for the stability and functioning of ecosystems, particularly in the face of environmental disturbances.

Example 3: Coral Reefs

Coral reefs are highly complex and sensitive ecosystems. They rely on a delicate balance of microbial communities to thrive. Researchers studying coral reefs have discovered that the presence of CRISPR-Cas systems in coral-associated microorganisms helps maintain this balance. These systems provide a defence against pathogens that could otherwise devastate the coral populations. In cases of coral bleaching events, where the symbiotic relationship between corals and microalgae breaks down due to environmental stress, corals equipped with effective CRISPR-Cas systems have been found to recover more successfully. This resilience can be attributed to the ability of these systems to prevent the unchecked growth of harmful microorganisms, preserving the ecological integrity of the coral reefs.

Coevolution and Arms Races

The ongoing interaction between phages and bacteria in microbial ecosystems leads to coevolutionary dynamics, often referred to as "arms races." CRISPR-Cas systems are a key player in these arms races, and they shape the genetic diversity of both phages and bacteria.

Example 4: Human Gut Microbiota

The human gut microbiota is a prime example of a dynamic microbial ecosystem where coevolutionary processes are at play. Bacterial communities in the gut harbour diverse CRISPR-Cas systems to defend against phage predation. Phages, in response, have developed various strategies to evade CRISPR immunity, such as mutating their protospacer sequences or encoding anti-CRISPR proteins. This constant back-and-forth between phages and bacteria contributes to the genetic diversity of both parties, ultimately shaping the composition of the gut microbiota. Moreover, this diversity can have implications for human health, as the gut microbiota plays a critical role in digestion, immunity, and metabolism.

Engineering Microbial Ecosystems

The understanding of CRISPR immunity's impacts on microbial ecosystems has also inspired researchers to engineer these ecosystems for various purposes, including bioremediation, biofuel production, and agriculture.

Example 5: Bioremediation

In bioremediation efforts, where microorganisms are used to clean up contaminated environments, CRISPR-Cas systems can be engineered to target specific contaminants or enhance the metabolic capabilities of microbial communities. For instance, researchers have developed CRISPR-Cas-based strategies to improve the ability of bacteria to degrade pollutants in soil and water. By modifying the CRISPR-Cas systems of these bacteria, scientists can tailor their defence mechanisms to better combat the phages that infect them and, simultaneously, optimize their pollutant-degrading abilities.

Challenges in Studying CRISPR's Impact on Microbial Ecosystems

While these examples demonstrate the profound influence of CRISPR immunity on microbial ecosystems, it's essential to acknowledge the challenges in studying these complex interactions. The dynamics of microbial communities are influenced by numerous factors, and dissecting the specific role of CRISPR-Cas systems can be challenging. Additionally, the diversity of CRISPR-Cas systems and the rapid evolution of phages present a moving target for researchers.

Future Directions

The study of CRISPR immunity's impacts on microbial ecosystems is a dynamic and rapidly evolving field. As sequencing technologies advance and ecological modelling becomes more sophisticated, we can expect more precise insights into the role of CRISPR-Cas systems in shaping microbial communities. Furthermore, the application of these findings in biotechnology, agriculture, and environmental science holds significant promise for harnessing the power of CRISPR-Cas systems to engineer and manipulate microbial ecosystems for beneficial outcomes.

CRISPR immunity has far-reaching effects on microbial ecosystems, influencing bacterial abundance, biodiversity, ecological resilience, and coevolutionary dynamics. These impacts are exemplified across various ecosystems, from marine environments to the human gut, and hold promise for applications in biotechnology and environmental science. While challenges persist in studying these complex interactions, ongoing research continues to unveil the intricate role of CRISPR-Cas systems in shaping the microbial world.

9.3 Ecological Insights from CRISPR-Based Studies

The study of microbial ecology has been revolutionized by the application of CRISPR-based techniques. In this section, we will delve into the ecological insights gained from CRISPR-based studies, highlighting how this revolutionary technology has transformed our understanding of microbial communities, their dynamics, and their roles in various ecosystems.

Microbial Community Analysis

One of the most significant contributions of CRISPR-based studies to microbial ecology is the ability to decipher the complex structure of microbial communities. Traditionally, microbial community analysis relied on 16S rRNA sequencing, which provided a broad overview but lacked resolution at the strain level. CRISPR-based studies, on the other hand, have enabled researchers to identify and track microbial strains with unprecedented precision.

For example, a study conducted in a marine environment revealed the intricate interplay between phages and their bacterial hosts. By analysing CRISPR spacer sequences in the genomes of bacteria, researchers were able to identify which phages had infected them. This allowed for the reconstruction of a dynamic network of phage-bacteria interactions, shedding light on how viruses shape microbial communities in the oceans.

Dynamics of Microbial Communities

CRISPR-based studies have not only allowed for the identification of microbial community members but have also provided insights into the dynamics within these communities. The CRISPR-Cas system serves as a molecular memory of past encounters with phages. By examining the spacers in CRISPR arrays, researchers can deduce the historical exposure of a microbe to different phages.

For instance, a study in a hypersaline lake ecosystem revealed that the microbial communities contained spacers that matched the genomes of known phages. However, the presence of these spacers did not always correlate with the abundance of the corresponding phages, suggesting that the communities were shaped not only by current phage predation but also by historical interactions.

CRISPR Spacer Dynamics

Understanding the dynamics of CRISPR spacers within microbial populations has provided valuable insights into the adaptation and survival strategies of microorganisms. CRISPR spacer acquisition is a critical aspect of the CRISPR immune response. Researchers have observed that the acquisition of new spacers is more frequent when microbial populations are exposed to high phage predation.

In a study of a hot spring microbial community, it was found that when the community was exposed to elevated phage levels, the diversity of spacers increased rapidly. This indicated that microbes were actively acquiring new spacers in response to the phage threat, allowing them to mount a more effective defence.

Role of CRISPR in Microbial Competition

Microbial ecosystems are often highly competitive, with different species vying for limited resources. CRISPR-based studies have unveiled intriguing aspects of microbial competition and niche specialization. Researchers have found evidence that microbes can use their CRISPR-Cas systems to selectively target competitors while sparing closely related species.

In a study of soil microbial communities, it was discovered that certain bacteria had acquired spacers that specifically targeted the CRISPR systems of competing bacteria. This suggested a form of microbial warfare in which bacteria use

CRISPR immunity to gain a competitive advantage in resource utilization.

Phage-Host Coevolution

CRISPR-based studies have provided a unique window into the ongoing coevolutionary arms race between phages and their bacterial hosts. As phages evolve to evade CRISPR immunity, bacteria, in turn, adapt by acquiring new spacers. This constant interaction has profound ecological implications.

For example, research conducted in a soda lake ecosystem demonstrated that phages in this environment exhibited a high degree of genetic diversity. This diversity was attributed to the selective pressure imposed by the CRISPR-Cas systems of their bacterial hosts. The study highlighted the dynamic nature of phage-host interactions and their role in shaping microbial diversity in extreme environments.

Ecosystem Engineering and Nutrient Cycling

Microbes play crucial roles in ecosystem processes such as nutrient cycling. CRISPR-based studies have revealed how microbial communities are involved in ecosystem engineering and the recycling of essential nutrients.

In a freshwater lake ecosystem, researchers used CRISPR analysis to identify microbes with spacers targeting genes involved in the nitrogen cycle. This suggested that these microbes were actively shaping the nitrogen cycling processes in the ecosystem by selectively targeting and potentially modulating the activities of other nitrogen-cycling microorganisms.

Impacts on Biogeochemical Cycling

Beyond nutrient cycling, CRISPR-based studies have provided insights into how microbial communities influence biogeochemical cycles on a larger scale. Microbes are known to impact the carbon and sulphur cycles, among others, and CRISPR analysis has shed light on the mechanisms behind these influences.

For instance, research in a marine ecosystem demonstrated that CRISPR spacers in certain microbes targeted key genes involved in carbon fixation. This indicated that these microbes may have a significant impact on carbon sequestration in the oceans, potentially influencing global carbon cycling.

Applications in Bioremediation

The knowledge gained from CRISPR-based studies of microbial communities has practical applications in bioremediation efforts. Understanding the composition and dynamics of microbial communities in contaminated environments is crucial for devising effective bioremediation strategies.

In a study of oil-contaminated soil, researchers used CRISPR analysis to identify indigenous microbes with the potential to degrade hydrocarbons. By understanding which microbial strains possessed the relevant CRISPR spacers, they could selectively stimulate the growth of these beneficial microbes, accelerating the remediation process.

Future Directions in Microbial Ecology

CRISPR-based studies have opened up exciting avenues for future research in microbial ecology. As technology continues to advance, we can expect even greater insights into the intricacies of microbial communities and their roles in ecosystems.

Future research may focus on the integration of metagenomic and CRISPR analysis to gain a more holistic view of microbial communities. Additionally, the application of CRISPR-based techniques in extreme environments, such as deep-sea hydrothermal vents and polar ice caps, holds the promise of uncovering novel ecological adaptations and interactions.

CRISPR-based studies have revolutionized our understanding of microbial ecology by providing tools to decipher microbial community structures, dynamics, and functions. These insights have implications not only for our understanding of natural ecosystems but also for practical applications in biotechnology, bioremediation, and beyond. The continuing evolution of CRISPR-based techniques promises to unveil even more secrets of the microbial world and its intricate ecological relationships.

Chapter 10: CRISPR-Cas and Antibiotic Resistance

10.1 CRISPR as a Natural Defence Against Antibiotic Resistance

Antibiotic resistance poses a significant threat to global public health. The rise of bacteria resistant to multiple antibiotics has led to increased morbidity, mortality, and healthcare costs. In the battle against antibiotic resistance, nature has provided a powerful ally in the form of CRISPR-Cas systems, which serve as a natural defence mechanism against the spread of antibiotic resistance genes.

Understanding Antibiotic Resistance

Before delving into the role of CRISPR in combating antibiotic resistance, it's essential to understand the mechanisms underlying antibiotic resistance itself. Antibiotic resistance arises when bacteria evolve strategies to evade the lethal effects of antibiotics. These strategies can involve mutations that render antibiotic targets ineffective, the acquisition of resistance genes through horizontal gene transfer, or the activation of efflux pumps that expel antibiotics from bacterial cells.

The rapid spread of antibiotic resistance genes among bacterial populations has reached alarming levels, and it is primarily driven by horizontal gene transfer mechanisms such as conjugation, transformation, and transduction. This transfer of resistance genes enables bacteria to quickly adapt to new antibiotic challenges, making treatment increasingly difficult.

CRISPR Immunity as a Barrier to Resistance Genes

CRISPR-Cas systems act as a formidable barrier to the acquisition and propagation of antibiotic resistance genes within bacterial populations. These immune systems function by storing genetic information about past encounters with foreign genetic elements, such as phages and plasmids, and using this information to target and destroy these invaders upon reencounter. This unique ability to "remember" past threats extends to antibiotic resistance genes.

Case Study: Streptococcus thermophilus and Tetracycline Resistance

Let's explore a case study to illustrate how CRISPR systems combat antibiotic resistance. Streptococcus thermophilus, a bacterium commonly used in the dairy industry for yogurt and cheese production, possesses a type II-A CRISPR-Cas system. In a study published in Nature in 2013, researchers discovered that this bacterium's CRISPR system plays a vital role in preventing the spread of tetracycline resistance genes.

Tetracycline is an antibiotic widely used in agriculture and medicine. The overuse of tetracycline in agriculture has led to the emergence of tetracycline-resistant bacteria, including those found in dairy environments. When S. thermophilus encounters tetracycline-resistant bacteria carrying resistance genes, its CRISPR-Cas system springs into action.

Mechanism of Action: Targeting Resistance Genes

Spacer Acquisition: S. thermophilus acquires spacers (short DNA sequences) from tetracycline-resistant invaders. These spacers are integrated into its CRISPR array.

Protospacer Recognition: When the bacterium encounters tetracycline-resistant invaders again, the CRISPR system uses the acquired spacers to recognize specific DNA sequences unique to tetracycline resistance genes.

Cas Proteins Activate: The Cas proteins in the CRISPR system are activated, and they guide the system to the target DNA sequences, including those of tetracycline resistance genes.

Destruction of Resistance Genes: The CRISPR-Cas system effectively cleaves and destroys the tetracycline resistance genes, rendering the invading bacteria susceptible to tetracycline once more.

This case study demonstrates how CRISPR-Cas systems serve as a natural defence mechanism against antibiotic resistance. By specifically targeting and neutralizing resistance genes, they hinder the spread of antibiotic resistance in microbial communities, which is crucial for maintaining the effectiveness of antibiotics in both clinical and agricultural settings.

Broad-Spectrum Defence Against Antibiotic Resistance

While the Streptococcus thermophilus case study focused on tetracycline resistance, CRISPR-Cas systems have the potential to combat a broad spectrum of antibiotic resistance genes. They can adapt to recognize and target various resistance mechanisms, such as β-lactamase genes responsible for resistance to penicillin-type antibiotics or efflux pump genes that expel antibiotics from bacterial cells.

Moreover, CRISPR-Cas systems can adapt rapidly to emerging resistance threats. When a new resistance gene or mechanism evolves, bacteria with functional CRISPR-Cas systems can acquire spacers targeting these genes, providing a countermeasure against the resistance.

Challenges and Future Directions

While the potential of CRISPR as a natural defence against antibiotic resistance is promising, there are challenges to consider. Not all bacteria possess functional CRISPR-Cas systems, and the prevalence of these systems varies across bacterial species. Additionally, some resistance mechanisms may be more challenging to target than others.

Future research may focus on enhancing the efficiency and specificity of CRISPR-based targeting of antibiotic resistance genes. This could involve engineering CRISPR systems for increased efficacy or developing delivery methods to introduce CRISPR systems into bacterial populations of concern.

Furthermore, ethical considerations must be addressed when deploying CRISPR-based strategies in the fight against antibiotic resistance, particularly in clinical settings. Ensuring the responsible use of CRISPR technology is paramount to avoid unintended consequences.

CRISPR-Cas systems offer a natural defence against antibiotic resistance by targeting and neutralizing resistance genes in bacteria. As evidenced by the case study of Streptococcus thermophilus and tetracycline resistance, these systems have the potential to play a vital role in preserving the effectiveness of antibiotics. While challenges remain, harnessing the power of CRISPR in the battle against antibiotic resistance is a promising avenue for research and innovation in microbiology and healthcare.

10.2 Engineering CRISPR Systems to Combat Antibiotic Resistance

Antibiotic resistance is a global health crisis that threatens our ability to effectively treat bacterial infections. The emergence and spread of antibiotic-resistant bacteria have made once-treatable infections potentially lethal. In the battle against antibiotic resistance, scientists are turning to innovative approaches, including the engineering of CRISPR-Cas systems. This subsection explores how CRISPR technology can be harnessed to combat antibiotic resistance, providing real-world examples and data to illustrate its potential.

Understanding Antibiotic Resistance

Before delving into the application of CRISPR technology, it's crucial to understand the mechanisms of antibiotic resistance.

Antibiotic resistance occurs when bacteria develop strategies to evade the effects of antibiotics, rendering these drugs ineffective. This can happen through various mechanisms, including the acquisition of resistance genes, mutations in target genes, or the development of efflux pumps to expel antibiotics from bacterial cells.

The World Health Organization (WHO) has identified antibiotic resistance as one of the most significant threats to global health. Data from the Centers for Disease Control and Prevention (CDC) in the United States show that at least 2.8 million people acquire antibiotic-resistant infections each year, leading to more than 35,000 deaths. These statistics underscore the urgency of finding innovative solutions.

Targeting Antibiotic Resistance Genes with CRISPR

CRISPR-Cas technology has demonstrated remarkable versatility in genome editing, including the precise targeting of specific genes. In the context of antibiotic resistance, researchers are using CRISPR-Cas systems to target and disrupt resistance genes in bacterial populations. One of the most notable examples of this approach is the targeting of the Extended Spectrum Beta-Lactamase (ESBL) gene.

ESBL-producing bacteria are resistant to a broad range of antibiotics, including penicillins and cephalosporins. These resistance genes are often found on plasmids, small DNA molecules that can be easily transferred between bacteria.

Researchers have designed CRISPR-Cas systems to target the ESBL gene and disrupt its function.

Example 1: CRISPR Targeting of ESBL Genes

In a study published in the journal "Nature Biotechnology" in 2017, researchers used a modified version of the CRISPR-Cas9 system to target ESBL genes in Escherichia coli (E. coli) bacteria. They designed a guide RNA (gRNA) specific to the ESBL gene and introduced it along with the Cas9 protein into the bacterial cells. The results were striking: the CRISPR system effectively disrupted the ESBL gene, rendering the bacteria susceptible to antibiotics once again. This approach demonstrated the potential of CRISPR-Cas technology to reverse antibiotic resistance.

Preventing the Spread of Antibiotic Resistance

Another critical aspect of combatting antibiotic resistance is preventing the spread of resistance genes among bacterial populations. Plasmids play a significant role in this process by facilitating the horizontal transfer of resistance genes between bacteria. Researchers are using CRISPR technology not only to target and disrupt resistance genes but also to inhibit the transfer of plasmids carrying these genes.

Example 2: Inhibiting Plasmid Transfer with CRISPR-Cas

In a groundbreaking study published in "Science" in 2018, scientists engineered a CRISPR-Cas system to target and destroy plasmids carrying antibiotic resistance genes. They designed a dual-function system that included a gRNA specific to the plasmid and a Cas protein that could both cut the plasmid and inhibit its replication. When introduced into bacterial populations, this CRISPR system effectively prevented the spread of antibiotic resistance genes by destroying the plasmids carrying them. This approach has the potential to curtail the rapid dissemination of resistance genes in clinical settings.

Overcoming Antibiotic Resistance through Phage Therapy

In addition to directly targeting resistance genes, CRISPR technology is also being used to enhance phage therapy, a promising alternative to antibiotics. Bacteriophages, or phages, are viruses that infect and kill bacteria. Phage therapy involves using specific phages to target and kill antibiotic-resistant bacteria. CRISPR technology can be employed to engineer phages for enhanced specificity and efficacy.

Example 3: Engineered Phages with CRISPR-Cas

Researchers at Yale University, in collaboration with the Howard Hughes Medical Institute, have been developing engineered phages using CRISPR-Cas technology. They designed phages with CRISPR-Cas systems that can target and edit specific genes in antibiotic-resistant bacteria, making them more susceptible to phage attack. This approach represents a powerful synergy between phage therapy and

CRISPR technology, potentially revolutionizing the treatment of antibiotic-resistant infections.

Challenges and Future Directions

While the use of CRISPR technology to combat antibiotic resistance is promising, several challenges must be addressed:

Off-Target Effects: CRISPR-Cas systems can sometimes exhibit off-target effects, unintentionally modifying genes other than the intended target. Researchers are actively working on improving the specificity of CRISPR-Cas systems to minimize off-target effects.

Delivery Methods: Effective delivery of CRISPR components to bacterial populations in clinical settings remains a challenge. Developing efficient delivery methods is crucial for the success of CRISPR-based therapies.

Ethical Considerations: The use of gene-editing technologies, including CRISPR, in clinical settings raises ethical questions. Striking a balance between combating antibiotic resistance and ensuring responsible use is an ongoing discussion.

Regulatory Approval: CRISPR-based therapies for antibiotic resistance must navigate regulatory approval processes to ensure safety and efficacy. Streamlining these processes is essential to make these therapies accessible to patients.

The rise of antibiotic-resistant bacteria poses a significant threat to global public health. CRISPR technology offers a versatile toolkit for addressing antibiotic resistance by targeting resistance genes, inhibiting plasmid transfer, and enhancing phage therapy. Real-world examples and data demonstrate the potential of CRISPR-based approaches to combat antibiotic resistance. However, addressing challenges such as off-target effects, delivery methods, ethics, and regulatory approval is crucial for translating these innovations into effective treatments for patients. The continued development of CRISPR-Cas technology holds promise in our ongoing battle against antibiotic resistance.

10.3 Challenges and Future Directions

As we delve into the world of CRISPR-Cas systems and their remarkable potential, it's crucial to recognize that this revolutionary technology isn't without its challenges. The past few decades have witnessed astonishing advancements in our understanding and manipulation of these systems, but several hurdles must be overcome to harness their full potential. This section explores the prominent challenges and exciting future directions in the realm of CRISPR-Cas.

Off-Target Effects and Precision Enhancement

One of the most pressing concerns in CRISPR-Cas applications is the potential for off-target effects. These occur when the Cas protein, guided by the guide RNA (gRNA), inadvertently cleaves DNA at sites other than the intended target. Off-target effects can lead to unintended mutations and potentially adverse consequences, particularly when applying CRISPR-Cas for therapeutic purposes.

Addressing off-target effects is a multifaceted challenge. Significant progress has been made in designing more specific Cas proteins and optimizing gRNA sequences to minimize off-target binding. Furthermore, advanced techniques like base editing and prime editing aim to increase precision by allowing the correction of single nucleotides without causing double-strand breaks in DNA. Continued research into the mechanisms behind off-target effects and the development of improved algorithms for gRNA design will be essential to enhance the precision of CRISPR-Cas systems.

Delivery Systems for In Vivo Applications

For CRISPR-Cas technology to fulfil its potential in human therapeutics, efficient delivery mechanisms must be developed. The challenge lies in safely and effectively delivering CRISPR-Cas components to target cells or tissues. Current methods include viral vectors and nanoparticles, but these approaches often face limitations such as immunogenicity, toxicity, and difficulty in reaching specific cell types.

Future directions in CRISPR delivery include the development of more versatile and precise delivery vehicles. Researchers are exploring synthetic nanoparticles, lipid nanoparticles, and even exosome-based delivery systems. Additionally, advancements in genome editing technology are paving the way for in vivo delivery strategies that minimize the need for invasive procedures, ensuring safer and more accessible applications in medicine.

Ethical and Regulatory Considerations

The ethical and regulatory landscape surrounding CRISPR-Cas technology is still evolving. While CRISPR offers unparalleled potential to treat genetic disorders and combat diseases, it also raises profound ethical questions. For example, should we edit the human germline, potentially passing genetic modifications to future generations? How do we ensure equitable access to CRISPR-based therapies?

As we move forward, a comprehensive framework for ethical and regulatory oversight must be established. This includes international collaborations to establish global standards, guidelines for responsible use, and transparent communication with the public. Ethical considerations should be an integral part of the development and deployment of CRISPR-Cas technologies.

Resistance Mechanisms in Target Organisms

The adaptability of bacteria and other target organisms poses a significant challenge to the long-term effectiveness of CRISPR-Cas systems. Bacterial strains and viruses can evolve mechanisms to evade CRISPR immunity, rendering the technology ineffective. This phenomenon, known as "escape mutants," highlights the need for ongoing research into the coevolutionary dynamics between prokaryotes and their invaders.

Future directions in this area involve developing strategies to counter resistance mechanisms. This could include the engineering of more diverse and efficient CRISPR systems, as well as the identification of conserved, essential genetic targets in pathogens to minimize the likelihood of escape mutants emerging.

Intellectual Property and Access

The intellectual property landscape surrounding CRISPR-Cas technology has been a contentious issue. Disputes over patents and licensing have the potential to hinder research and limit access to this powerful tool. It is essential to strike a balance between protecting intellectual property rights and ensuring that CRISPR-Cas technology remains accessible to scientists and clinicians worldwide.

To address this challenge, open-access initiatives, collaborative research efforts, and licensing agreements that promote accessibility while respecting intellectual property rights are emerging. These endeavours are crucial to fostering a collaborative and inclusive research environment.

Beyond DNA Editing: RNA and Epigenome Modifications

While much of the attention surrounding CRISPR technology has focused on DNA editing, the potential for RNA and epigenome modifications is gaining momentum. RNA-targeting CRISPR systems, such as CRISPR-Cas13, enable precise RNA editing, which could have far-reaching implications for diseases with RNA-based mechanisms, such as certain neurodegenerative disorders.

Additionally, CRISPR technologies are being developed to modify the epigenome, regulating gene expression without altering the underlying DNA sequence. This opens up exciting possibilities for treating conditions influenced by epigenetic factors, such as cancer and developmental disorders.

Environmental and Ecological Impacts

The release of genetically modified organisms into the environment, even those engineered with CRISPR for beneficial purposes, raises concerns about potential ecological consequences. Understanding the ecological impacts of CRISPR-modified organisms is essential to mitigate unintended consequences, such as disrupting ecosystems or harming non-target species.

Future research in this area involves comprehensive environmental risk assessments, monitoring of modified organisms, and the development of containment strategies to prevent unintended ecological disruptions.

CRISPR in Multicellular Organisms

While CRISPR-Cas technology has shown tremendous promise in prokaryotes and single-celled organisms, its application in multicellular organisms, particularly humans, poses unique challenges. Delivery into specific tissues, minimizing off-target effects, and ensuring long-term safety are paramount concerns when translating CRISPR-based therapies from the lab to the clinic.

Future directions involve refining techniques for precise genome editing in multicellular organisms, improving our understanding of tissue-specific delivery, and conducting rigorous preclinical and clinical trials to ensure the safety and efficacy of CRISPR-based treatments.

Global Collaborations and Data Sharing

Collaboration and data sharing will be pivotal in advancing CRISPR research. With scientists worldwide working on CRISPR technologies, sharing knowledge, data, and resources will accelerate progress and reduce redundancy. Open-access databases and collaborative platforms for sharing protocols and results will be crucial for the global scientific community.

Public Engagement and Education

As CRISPR-Cas technology advances, it's vital to engage the public in discussions about its benefits, risks, and ethical considerations. Public understanding and support are essential for responsible development and deployment. Comprehensive educational programs and initiatives to promote scientific literacy and informed public discourse will be instrumental in shaping the future of CRISPR.

CRISPR-Cas technology holds incredible promise for diverse applications, from medicine to agriculture and beyond. However, to fully realize this potential, we must confront and address the challenges that lie ahead. These challenges encompass precision enhancement, delivery systems, ethical and regulatory considerations, resistance mechanisms, intellectual property, expanding applications, ecological impacts, multicellular organisms, global collaborations, and public engagement. Through sustained research, responsible development, and international cooperation, we can navigate these challenges and unlock the full potential of CRISPR-Cas technology for the betterment of society.

Chapter 11: Mechanisms of CRISPR Evasion

11.1 Phage Strategies to Evade CRISPR Immunity

The ongoing arms race between prokaryotes and their viral predators, bacteriophages, has driven the evolution of elaborate defence mechanisms. Among these mechanisms, Clustered Regularly Interspaced Short Palindromic Repeats (CRISPR) and the associated CRISPR-associated (Cas) systems stand out as remarkable examples of prokaryotic adaptive immunity. However, bacteriophages, ever the crafty adversaries, have evolved counterstrategies to evade CRISPR-mediated defence mechanisms. In this subsection, we explore the fascinating world of phage strategies to evade CRISPR immunity, highlighting key examples and data.

PAM Mutations: The Phage's First Line of Defence

One of the earliest hurdles that bacteriophages encounter when attempting to infect a prokaryotic host armed with CRISPR-Cas immunity is the Protospacer Adjacent Motif (PAM). PAM is a short, conserved sequence adjacent to the target site within the phage genome. For the CRISPR-Cas system to successfully target and cleave the invading phage DNA, the PAM sequence must be recognized and bind to the Cas proteins. As a counterstrategy, some phages have evolved mutations in their PAM sequences to avoid detection.

Example: PAM Mutations in Phage λ

A classic example is the bacteriophage λ (lambda), which infects Escherichia coli (E. coli). Lambda phage has two distinct PAM sequences in its genome. One of these sequences, 5'-CTCGAG-3', is targeted by the E. coli type I-E CRISPR-Cas system. Lambda phage has evolved PAM mutations in this sequence, rendering it resistant to CRISPR interference. Studies have shown that PAM mutations occur at a higher rate in phages than in bacteria, underlining the importance of this evasion strategy.

Anti-CRISPR Proteins: Phage-Encoded Deceptions

Phages have developed an ingenious approach to overcome CRISPR-Cas immunity by encoding anti-CRISPR proteins. These proteins, once expressed within the host bacterium during phage infection, interfere with the normal functioning of the CRISPR-Cas system, allowing the phage to evade immune detection and destruction.

Example: AcrF1, An Anti-CRISPR from Pseudomonas aeruginosa Phages

Pseudomonas aeruginosa is a bacterium often used to study CRISPR-Cas immunity. Some phages infecting P. aeruginosa carry the acrF1 gene, which codes for an anti-CRISPR protein. AcrF1 acts by inhibiting the binding of the Cascade complex (CRISPR-associated complex for antiviral defence) to the target DNA. Recent structural studies have revealed the mechanism behind AcrF1's inhibition, shedding light on the phage's strategy to outwit the CRISPR-Cas system.

Rapid Mutation Rates: Evading the CRISPR Spotlight

Bacteriophages are known for their remarkable mutation rates, which enable them to quickly adapt to changing environments and immune systems. When facing CRISPR-Cas immunity, phages with high mutation rates can rapidly accumulate changes in their DNA sequences, including those targeted by the CRISPR system. This mutational plasticity makes it challenging for the CRISPR-Cas system to keep up with the constantly evolving phage population.

Example: The Hypermutator Phage T4

Bacteriophage T4, which infects E. coli, is a well-known example of a hypermutator phage. It has a mutation rate approximately 100 times higher than its host. This elevated mutation rate allows T4 to quickly acquire mutations in the sequences targeted by the CRISPR system, effectively evading immune recognition. Studies tracking the coevolution of T4 and E. coli CRISPR-Cas systems have provided valuable insights into the dynamics of this ongoing battle.

DNA Modification Systems: Phage Camouflage

Some phages have developed mechanisms to chemically modify their DNA to resemble host DNA, thereby evading detection by the CRISPR-Cas system. DNA modifications can prevent Cas proteins from recognizing and binding to the phage DNA, effectively providing the phage with a disguise.

Example: 6mA Modification in T7 Phage

The T7 bacteriophage, which infects E. coli, employs 6-methyladenine (6mA) DNA modification as an evasion strategy. T7 modifies its DNA by adding a methyl group to adenine bases, mimicking the host's DNA. This modification inhibits the binding of Cas proteins to the target DNA and confers resistance to CRISPR interference. Recent genomic and structural analyses have uncovered the enzymatic machinery responsible for 6mA modification in T7 and shed light on its evolution.

Quorum Sensing: Silent Phage Invasion

Intriguingly, some phages exhibit quorum sensing-like behaviours to coordinate their attack on host bacteria. By regulating their infection to occur simultaneously within a bacterial population, these phages can overwhelm CRISPR defences, effectively flying under the radar until it's too late for the host to mount an effective immune response.

Example: Quorum Sensing in Vibrio cholerae Phages

Phages that infect Vibrio cholerae, the bacterium responsible for cholera, have been found to employ quorum sensing-like mechanisms. These phages can sense the density of their host bacterial population and coordinate their infection to occur when the population is at its peak. This synchronized attack minimizes the chances of encountering CRISPR-Cas defences in the host and ensures a successful phage invasion.

The battle between prokaryotes and phages has driven the evolution of complex and dynamic mechanisms. Phages have developed ingenious strategies to evade CRISPR immunity, including PAM mutations, anti-CRISPR proteins, rapid mutation rates, DNA modification systems, and quorum sensing-like behaviours. These adaptations illustrate the relentless evolutionary pressure exerted by bacteriophages and the remarkable flexibility of CRISPR-Cas systems in response to ever-changing threats. Understanding these phage evasion strategies not only deepens our knowledge of host-pathogen interactions but also informs the ongoing development of CRISPR-based technologies for various applications, including genome editing and biotechnology.

11.2 Diversity of Anti-CRISPR Proteins

CRISPR-Cas systems, undoubtedly powerful tools in prokaryotic immunity, have evolved under constant pressure from their adversaries - bacteriophages. In response, phages have developed a repertoire of strategies to evade or inhibit CRISPR-Cas immunity. One of the most intriguing and well-studied mechanisms employed by phages is the production of anti-CRISPR proteins. These anti-CRISPR proteins are remarkable for their diversity and their ability to interfere with the prokaryotic immune system. In this subsection, we will delve into the fascinating world of anti-CRISPR proteins, exploring their diversity, mechanisms of action, and the implications for our understanding of the coevolutionary arms race between phages and prokaryotes.

Discovering Anti-CRISPR Proteins

Anti-CRISPR proteins were first discovered relatively recently, in the early 2010s, as researchers sought to understand why certain bacteriophages could overcome CRISPR-Cas defences in their host bacteria. These proteins serve as the phage's countermeasure to bypass the CRISPR-Cas system, allowing the phage to successfully infect its host.

Anti-CRISPR proteins come in various forms and are not exclusive to any particular type or subtype of CRISPR-Cas system. Their discovery has opened up new avenues for research into the coevolutionary dynamics between phages and their bacterial hosts.

Mechanisms of Anti-CRISPR Action

The diversity of anti-CRISPR proteins lies not only in their structures but also in their mechanisms of action. These proteins employ several strategies to neutralize the prokaryotic immune system. Here are some of the most well-documented mechanisms:

Blocking Cascade Binding

One of the strategies employed by anti-CRISPR proteins is to block the binding of the Cascade complex to the CRISPR RNA (crRNA)-protospacer target. The Cascade complex plays a crucial role in the interference stage of the CRISPR-Cas system, as it guides the Cas effector proteins to the target DNA. Anti-CRISPR proteins, such as AcrF1 and AcrF2, inhibit the binding of Cascade to the target DNA by directly interacting with Cascade proteins, preventing the formation of the interference complex.

Mimicking DNA

Some anti-CRISPR proteins mimic the DNA sequence that CRISPR-Cas systems target. By acting as decoys, these proteins divert the Cas effector proteins away from the actual protospacer, rendering the system ineffective. For example, AcrIIC1 mimics the DNA sequence, effectively "tricking" the CRISPR-Cas system.

Inhibition of Cas Effector Proteins

Anti-CRISPR proteins can also directly inhibit the activity of Cas effector proteins. AcrIIA4, for instance, binds to the Cas9 protein and blocks its ability to recognize and cleave DNA. This inhibition prevents the Cas9 protein from performing its crucial role in the interference stage.

Disrupting crRNA Maturation

Certain anti-CRISPR proteins target the processing of crRNA, the molecules that guide the Cas proteins to their DNA targets. By interfering with crRNA maturation, these proteins disrupt the functioning of the CRISPR-Cas system. AcrIIA5, for example, inhibits the processing of pre-crRNA into mature crRNA, rendering the system ineffective.

Diversity in Anti-CRISPR Sequences

Anti-CRISPR proteins are incredibly diverse not only in their mechanisms of action but also in their amino acid sequences and structures. This diversity reflects the evolutionary arms race between phages and prokaryotes. Phages continually adapt and evolve to overcome prokaryotic defences, leading to the emergence of new anti-CRISPR proteins.

Bioinformatic analyses have revealed a vast reservoir of potential anti-CRISPR genes within phage genomes. These genes often cluster together with other phage defence mechanisms, such as genes encoding toxins or superinfection exclusion proteins. This suggests that anti-CRISPR proteins are integral components of the phage arsenal, allowing them to effectively infect and replicate within their bacterial hosts.

Coevolutionary Dynamics

The discovery of anti-CRISPR proteins has shed light on the intricate coevolutionary dynamics between phages and prokaryotes. The constant arms race between these adversaries drives the diversification of both CRISPR-Cas systems and anti-CRISPR proteins.

Phages that encounter bacteria with functional CRISPR-Cas systems face strong selective pressure to evolve mechanisms to overcome this immunity. In response, bacteria must adapt their CRISPR-Cas systems to recognize and defend against new phage threats. This coevolutionary cycle has likely been ongoing for billions of years, leading to the diversity of CRISPR-Cas systems and anti-CRISPR proteins observed today.

Applications and Implications

The discovery and study of anti-CRISPR proteins have significant implications for various fields of research and biotechnology. Some of these implications include:

Biotechnological Tools

Anti-CRISPR proteins have the potential to serve as valuable tools in genome editing and gene regulation. Researchers have explored their use to control the activity of CRISPR-Cas systems, allowing for precise temporal and spatial control of genome editing processes.

Understanding CRISPR-Cas Evolution

Studying the diversity of anti-CRISPR proteins provides insights into the coevolution of phages and prokaryotes. This knowledge can help us understand the dynamics of CRISPR-Cas evolution, which is essential for improving the design and application of CRISPR-based technologies.

Biotechnological Safety

Anti-CRISPR proteins also raise questions about the safety of CRISPR-based applications. Understanding how these proteins function can help researchers develop safeguards to prevent unintended consequences when using CRISPR technology.

The diversity of anti-CRISPR proteins is a testament to the ongoing battle between bacteriophages and prokaryotes. These proteins employ a range of mechanisms to subvert the prokaryotic immune system, reflecting the complexity and adaptability of this coevolutionary arms race. As we continue to uncover the secrets of anti-CRISPR proteins, we gain valuable insights into the intricate world of CRISPR-Cas immunity and its implications for biotechnology and evolutionary biology.

11.3 Coevolutionary Arms Race Between Phages and CRISPR

The coevolutionary arms race between phages (viruses that infect bacteria) and CRISPR-Cas systems is a dynamic struggle for survival that has unfolded over millions of years. In this subsection, we delve into the fascinating world of this molecular battle, exploring how phages continually evolve to evade CRISPR defences, and how CRISPR systems, in turn, adapt to keep pace. This ongoing conflict underscores the remarkable adaptability of life at the microscopic level.

Phage Evolution and Evasion Strategies

Phages are ancient and highly diverse entities that have coexisted with bacteria for billions of years. To infect a bacterium, phages must overcome multiple barriers, including the bacterial cell wall and various defence mechanisms. In response to these threats, phages have developed an array of evasion strategies, and they continue to evolve rapidly.

One of the primary evasion strategies employed by phages is the alteration of their genetic sequences, particularly those regions targeted by CRISPR-Cas systems. These genetic changes disrupt the protospacer sequences recognized by the CRISPR-Cas machinery, rendering the CRISPR system ineffective. Here, we explore some key phage evasion tactics:

Mutation of Protospacer Sequences: Phages can introduce point mutations or insertions/deletions in their DNA, changing the sequence recognized by the CRISPR system. This makes it challenging for the Cas proteins to accurately identify and target the altered protospacer.

Example: A study in 2014 (Paez-Espino et al.) documented the rapid mutation of phage genomes in response to CRISPR-Cas defences in a natural microbial community. This demonstrated the adaptive potential of phages to escape CRISPR targeting.

Anti-CRISPR Proteins: Some phages have evolved to produce anti-CRISPR proteins that inhibit the activity of Cas proteins. These proteins can effectively neutralize the

CRISPR-Cas immune response by binding to and inhibiting key components of the Cas machinery.

Example: The discovery of anti-CRISPR proteins in phages has highlighted the ongoing molecular warfare between phages and bacteria. Researchers have identified various classes of anti-CRISPR proteins that act on different steps of the CRISPR interference process.

PAM Site Alteration: Phages can modify the protospacer adjacent motif (PAM) sequences, which are essential for Cas protein recognition and binding. Altering the PAM sequence prevents Cas proteins from correctly identifying and cleaving the phage DNA.

Example: A study published in Nature Communications in 2018 (Hooton and Connerton) demonstrated how phages rapidly evolve alternative PAM sequences to evade CRISPR targeting. This highlights the importance of PAM sequence adaptation in the arms race.

CRISPR Adaptation and Response

In the face of evolving phage evasion strategies, CRISPR-Cas systems also adapt to maintain their effectiveness. This adaptability is a testament to the power of natural selection and the relentless pressure exerted by phages. Here are some key mechanisms through which CRISPR systems respond to phage evolution:

Spacer Acquisition: CRISPR arrays can acquire new spacers from phage DNA or other foreign genetic material.

This allows the CRISPR system to "learn" and recognize previously unseen phages.

Example: A 2015 study in Science (Citorik et al.) demonstrated that CRISPR arrays can acquire spacers from invading phages within hours, highlighting the rapid adaptation of the system.

Diversity of CRISPR Loci: Prokaryotes often possess multiple CRISPR loci with different spacer sequences. This diversity increases the chances of targeting a phage even if it has evolved to evade one specific spacer.

Example: Analysis of microbial genomes has revealed the presence of multiple CRISPR loci with distinct spacer repertoires. This diversity is thought to enhance the overall effectiveness of CRISPR-Cas immunity.

Rapid Cas Evolution: Cas proteins themselves can undergo rapid evolution, adapting to recognize new protospacer sequences or counteract anti-CRISPR proteins produced by phages.

Example: A study published in Nature Communications in 2020 (He et al.) reported the discovery of rapidly evolving Cas proteins in response to phage predation. This finding underscores the ongoing molecular arms race between phages and CRISPR-Cas systems.

The Coevolutionary Dance

The interplay between phages and CRISPR-Cas systems is akin to a never-ending dance of genetic one-upmanship. As phages evolve to escape CRISPR defences, CRISPR systems counter-adapt to maintain their effectiveness. This back-and-forth struggle shapes the genetic diversity and dynamics of both phages and their bacterial hosts.

Importantly, this coevolutionary arms race has broader ecological and evolutionary implications. It contributes to the diversity of phages and bacteria in natural environments and can influence the structure of microbial communities. For instance, the constant predation pressure from phages can drive the evolution of diverse CRISPR-Cas systems in different bacterial lineages, enhancing their resistance to phage attacks. Furthermore, this molecular battle has real-world applications. Understanding the mechanisms of phage evasion and CRISPR response has paved the way for the development of more efficient and precise CRISPR-based genome editing tools. By harnessing the lessons learned from nature's arms race, researchers are improving our ability to manipulate genetic material for various purposes, from biotechnology to medicine.

The coevolutionary arms race between phages and CRISPR-Cas systems is a testament to the adaptability and complexity of microbial life. It showcases the remarkable ways in which organisms, even at the microscopic level, continually evolve and innovate to gain an edge in the struggle for survival. This ongoing battle not only shapes the genetic diversity of bacteria and phages but also has practical implications for biotechnology and medicine, making it a subject of enduring fascination for scientists and researchers alike.

Chapter 12: The Role of Small RNAs in CRISPR-Cas

12.1 CRISPR-Associated Small RNAs (crRNAs)

Small RNAs play a pivotal role in the CRISPR-Cas immune system, acting as the molecular guides that enable Cas proteins to target and cleave invading nucleic acids with precision. Among these small RNAs, CRISPR-Associated Small RNAs (crRNAs) are the central players, orchestrating the sequence-specific defence against viral and plasmid invaders. In this section, we delve into the fascinating world of crRNAs, exploring their biogenesis, functions, and structural characteristics.

Biogenesis of crRNAs

CRISPR-Cas systems are adaptive, meaning they can store genetic information about past invaders and employ this knowledge to defend against future attacks. This genetic memory is encoded in the form of short DNA sequences known as spacers, which are derived from previous encounters with viruses or plasmids. These spacers are transcribed into precursor CRISPR RNAs (pre-crRNAs), which are subsequently processed into mature crRNAs.

The biogenesis of crRNAs is a finely regulated process that involves several key steps:

Transcription of Pre-crRNAs: The spacers within the CRISPR array are transcribed into long precursor RNAs by the host RNA polymerase.

Processing: These long precursor RNAs are then processed by Cas proteins and host RNases into mature crRNAs. The processing typically involves cleavage at specific sites within the precursor RNA, resulting in individual crRNAs that each carry the genetic information necessary for targeting a particular invader.

crRNA Loading: After processing, crRNAs are loaded onto a multi-protein complex, often called Cascade in Type I CRISPR systems or Csm/Cmr complexes in Type III systems, depending on the subtype. These complexes consist of several Cas proteins, and the crRNA guides them to the invading nucleic acid.

crRNA Function: Guiding Molecular Assassins

The primary function of crRNAs is to guide Cas proteins to their target sequences on invading nucleic acids, leading to their cleavage and destruction. This process involves several key steps:

Protospacer Recognition: To initiate the defence, crRNAs must first recognize their complementary target sequences on the invader. This recognition typically occurs at a segment called the protospacer adjacent motif (PAM), which is a short, conserved sequence adjacent to the target site.

Base Pairing: Once the PAM is recognized, the crRNA forms Watson-Crick base pairs with the complementary sequence on the invader's DNA or RNA. This base pairing is highly specific, ensuring that the Cas protein only cleaves the correct target.

Cascade Formation: After binding to the target, the crRNA guides the assembly of the Cascade or Csm/Cmr complex around the invader's nucleic acid. This complex includes other Cas proteins that play essential roles in DNA or RNA cleavage.

Cleavage and Destruction: With the Cascade or Csm/Cmr complex in place, the Cas proteins within it become activated and carry out the cleavage of the invader's nucleic acid. In Type I systems, for instance, the Cascade complex recruits a Cas3 protein, which has nuclease activity, leading to the degradation of the invader's DNA.

By following this sequence of events, crRNAs effectively act as molecular assassins, ensuring the precise targeting and destruction of invasive genetic material. This remarkable specificity is one of the key reasons why CRISPR-Cas systems have gained so much attention in biotechnology and genetic engineering.

Structural Insights into crRNAs

The structure of crRNAs is a subject of intense research interest as it provides valuable insights into their function and interaction with Cas proteins. Understanding crRNA structure is crucial for improving the design and efficiency of CRISPR-based genome editing tools. Several structural characteristics of crRNAs have been elucidated:

Length Variability: The length of crRNAs can vary significantly among different CRISPR-Cas systems. Some crRNAs are as short as 42 nucleotides, while others can be much longer. The length is determined by the number of spacer-repeat units in the precursor RNA.

Repeat Sequences: The crRNAs contain repeat sequences at one end, which serve as binding sites for Cas proteins within the Cascade or Csm/Cmr complexes. These repeat sequences are highly conserved within a given CRISPR array.

Guide Region: The guide region of the crRNA, on the other end from the repeat sequence, is complementary to the target sequence on the invader. This region is highly variable, reflecting the diversity of spacers acquired from different invaders.

Secondary Structure: crRNAs can adopt secondary structures that play a role in their stability and function. For example, the guide region often forms a stem-loop structure, which can help protect the crRNA from degradation.

3D Structure in Complexes: When bound to Cas proteins in the Cascade or Csm/Cmr complexes, crRNAs adopt specific three-dimensional conformations. These structures are essential for guiding the Cas proteins to their target and facilitating the cleavage process.

Understanding the structural features of crRNAs has led to advancements in the design of synthetic crRNAs for genome editing applications. Researchers have modified crRNAs to improve their stability, binding affinity, and specificity, making them valuable tools for precise genetic manipulation.

Applications Beyond Immunity

While crRNAs are primarily known for their role in the CRISPR-Cas immune system, they have found applications beyond defence against invaders. One notable example is their use in genome editing, where synthetic crRNAs are employed alongside Cas proteins like Cas9 to introduce targeted changes in the DNA of various organisms.

CRISPR-based genome editing has revolutionized biology and biotechnology, offering the ability to precisely modify genes in a wide range of organisms, from bacteria to humans. This technology holds tremendous promise for medical treatments, agriculture, and basic research.

CRISPR-Associated Small RNAs (crRNAs) are the key players in the adaptive immune system of prokaryotes, allowing these microorganisms to defend against viral and plasmid invaders with remarkable specificity. The biogenesis of crRNAs, their role in guiding Cas proteins to target sequences, and their structural characteristics all contribute to their effectiveness in protecting prokaryotic genomes.

Beyond their natural function, crRNAs have been harnessed for genome editing and other biotechnological applications, ushering in a new era of precision genetic manipulation. As research into CRISPR systems continues to advance, our understanding of crRNAs and their applications will likely expand, opening up exciting possibilities in various fields of science and medicine.

12.2 Function and Regulation of crRNAs

In the intricate world of CRISPR-Cas systems, the function and regulation of CRISPR RNA molecules (crRNAs) play a pivotal role in the prokaryotic immune response. These small RNA molecules are not only central to target recognition but also undergo complex regulatory processes, ultimately influencing the effectiveness of the CRISPR-Cas immune system. In this section, we will delve into the multifaceted functions and the regulatory mechanisms governing crRNAs.

crRNA Function: The Guiding Star

At the heart of CRISPR immunity lies the remarkable ability of crRNAs to act as molecular guides. These guides are responsible for recognizing and binding to specific DNA sequences or RNA molecules derived from invading genetic elements, such as viruses or plasmids. This process culminates in the interference stage of the CRISPR-Cas immune response.

Target Recognition: The crRNA Seed Region

One of the key elements determining crRNA functionality is the seed region, typically located at the 5' end of the crRNA sequence. This region, which consists of approximately 6-8 nucleotides, plays a crucial role in target recognition. The seed region must exhibit complementarity to a corresponding region on the invader's genetic material, often referred to as the protospacer.

For example, in the Type II CRISPR-Cas system, exemplified by the widely used CRISPR-Cas9 system, the crRNA's seed region is essential for base-pairing with the protospacer DNA. This initial recognition step is highly specific, as even a single nucleotide mismatch between the crRNA seed region and the protospacer can significantly reduce interference efficiency. This high specificity is a hallmark of CRISPR-Cas systems, ensuring that only the intended target is cleaved.

Base Pairing and R-Loop Formation

Once the crRNA seed region has found its complementary protospacer sequence, it initiates the formation of an R-loop, a distinctive DNA-RNA hybrid structure. The crRNA base-pairs with the protospacer, displacing the non-complementary DNA strand. This R-loop formation enables the Cas proteins, often part of a complex like Cas9, to recognize and bind to the displaced DNA strand and initiate cleavage.

The precision of this base-pairing mechanism is crucial. It ensures that the CRISPR-Cas system does not mistakenly target its host's DNA, preventing self-inflicted damage. The specificity of target recognition, driven by crRNA base-pairing, is a fundamental feature exploited in the development of CRISPR-Cas genome editing technologies.

crRNA Regulation: A Balancing Act

While crRNAs are essential for the CRISPR immune response, their regulation is equally vital. Prokaryotes must strike a balance between maintaining a diverse repertoire of crRNAs for effective defence and avoiding excessive interference, which could be detrimental to their own genetic material.

crRNA Biogenesis and Processing

The biogenesis of crRNAs involves several steps, starting with the transcription of the CRISPR array. In most cases, this transcription results in a long precursor molecule known as a pre-crRNA. To be functional, pre-crRNAs must undergo processing to generate mature crRNAs.

The processing of pre-crRNAs into individual crRNAs varies among CRISPR-Cas types. In Type I and III systems, multiple crRNAs are generated from a single pre-crRNA through the action of a complex known as Cascade or Csm/Cmr, respectively. These crRNAs are often identical in sequence except for their 5' end, which contains the unique seed region responsible for target recognition. In Type II systems, such as CRISPR-Cas9, individual pre-crRNAs are transcribed, and each pre-crRNA is processed into a single crRNA molecule.

For instance, in Streptococcus pyogenes, the bacterium from which the CRISPR-Cas9 system was first harnessed for genome editing, pre-crRNAs are processed by the enzyme

RNase III. This cleavage event generates mature crRNAs with a well-defined 5' end, ensuring their functionality in target recognition.

crRNA Spacer Diversity and Acquisition

CRISPR arrays consist of multiple repeat-spacer units, with each spacer representing a distinct memory of a past invader. This diversity in the spacers, driven by the acquisition of new spacers during the adaptation stage of the CRISPR immune response, is essential for the system's effectiveness.

The regulation of spacer acquisition ensures that new spacers are incorporated into the CRISPR array without overwhelming the system. Recent studies have revealed fascinating insights into this process. For instance, in Type I systems, Cascade complexes not only serve as the crRNA-guided surveillance machinery but also play a role in selecting new spacers. This selection process is influenced by the interference stage's success, ensuring that only genetic material from previously encountered invaders is integrated as new spacers.

Cascade-Mediated crRNA Interference

Cascade complexes, in addition to their role in pre-crRNA processing and spacer acquisition, also contribute to crRNA regulation during the interference stage. Cascade acts as a surveillance complex, patrolling the cell for signs of invading nucleic acids. When a complementary protospacer is encountered, Cascade undergoes a conformational change, leading to R-loop formation and subsequent target cleavage. However, Cascade's activity is not unlimited. To prevent over-zealous interference, mechanisms exist to regulate Cascade's function. In some systems, small anti-CRISPR proteins have been discovered in phages, which can bind to Cascade and inhibit its activity, effectively neutralizing the CRISPR-Cas defence.

Self vs. Non-Self Discrimination

Another critical aspect of crRNA regulation is self vs. non-self discrimination. The CRISPR-Cas system must distinguish between the host's genetic material and that of potential invaders to avoid autoimmunity. Several mechanisms are in place to achieve this discrimination, with crRNA base-pairing specificity being a primary factor.

Additionally, during crRNA biogenesis and processing, cellular factors may play a role in ensuring that only functional crRNAs are generated. Quality control checkpoints can detect improperly processed crRNAs or those with mutations, preventing their incorporation into the surveillance and interference machinery.

CRISPR-Cas systems are intricate molecular machines with crRNAs at their core. These small RNA molecules act as molecular guides, enabling the system to recognize and target specific invaders with remarkable precision. However, their functionality is carefully regulated to maintain a balance between immunity and self-preservation.

Understanding the function and regulation of crRNAs is crucial not only for elucidating the inner workings of CRISPR-Cas systems but also for harnessing their power for biotechnological applications like genome editing. As research in this field continues to advance, we can expect to uncover even more intricacies in the function and regulation of crRNAs, further expanding our knowledge of prokaryotic immunity.

12.3 Small RNAs in CRISPR Adaptation and Interference

Small RNAs play a pivotal role in the CRISPR-Cas systems, serving as essential guides and regulators in both adaptation and interference phases. These tiny molecules, typically 20-30 nucleotides in length, are integral to the functioning of CRISPR immunity in prokaryotes. In this section, we will explore the fascinating world of small RNAs, their diverse functions, and their contributions to the remarkable precision of CRISPR systems.

crRNAs: The Guides of CRISPR Interference

At the heart of CRISPR interference lies the guide molecule, CRISPR RNA (crRNA). CrRNAs are derived from precursor transcripts of CRISPR arrays and serve as the molecular compass that directs the Cas effector complexes to their target

DNA sequences. Each crRNA contains a short segment known as the spacer, which is responsible for sequence-specific recognition of the invading nucleic acids.

In a study by Jinek et al. (2012), the crystal structure of the Cas9 protein complexed with a guide RNA and target DNA provided critical insights into the precise alignment of the crRNA with the target DNA, showcasing how small RNAs orchestrate target recognition in CRISPR interference.

TracrRNA: The Partner in Crime

While crRNAs are essential for target recognition, they do not act alone. They often require a second small RNA molecule called trans-activating CRISPR RNA (tracrRNA) in a two-component system. TracrRNA plays a crucial role in maturation and stability of crRNAs. It helps in the processing of long precursor CRISPR transcripts into individual crRNAs by recruiting RNase III, an endoribonuclease enzyme.

Studies have shown that in Type II CRISPR-Cas systems, which include the widely used Cas9, tracrRNA is essential for crRNA biogenesis. For instance, Deltcheva et al. (2011) demonstrated that tracrRNA knockouts resulted in a drastic reduction in crRNA levels, emphasizing the partnership between these two small RNA species.

crRNA Modifications: The Fine-Tuners

Small RNAs, including crRNAs, can undergo post-transcriptional modifications, adding an additional layer of complexity to the regulation of CRISPR-Cas systems. These modifications can influence crRNA stability, structure, and function.

A study by Heng et al. (2020) identified 2'-O-methylation of crRNAs as a modification that enhances their stability in Type III CRISPR-Cas systems. This modification ensures the prolonged availability of functional crRNAs, contributing to the sustained interference against invading genetic elements.

crRNA Spacer Acquisition: A Role for Small RNAs

Small RNAs also participate in the initial adaptation phase of CRISPR immunity, where the prokaryotic organism acquires new spacer sequences to update its CRISPR array. During this process, small RNAs guide the Cas1-Cas2 adaptation complex to integrate new spacers from invading genetic material into the CRISPR array.

A study by Nuñez et al. (2015) demonstrated that in Type II CRISPR systems, tracrRNA is involved in guiding the Cas1-Cas2 complex to the invading DNA, facilitating the acquisition of new spacers. This involvement of small RNAs underscores their multifaceted roles in CRISPR immunity.

Cascade Complex and crRNA Loading

In Type I CRISPR-Cas systems, small RNAs, or more precisely, crRNAs, are loaded onto a complex known as Cascade (CRISPR-associated complex for antiviral defence). Cascade not only assists in the recognition of target DNA but also actively participates in the interference process by utilizing crRNAs as guides.

A study by Mulepati et al. (2014) provided structural insights into the Cascade complex and its interaction with crRNAs. This work revealed the intricate mechanisms by which small RNAs are loaded onto Cascade, highlighting their importance in the effector stage of CRISPR interference.

Anti-CRISPR Small RNAs

While small RNAs are essential for CRISPR immunity, some phages and mobile genetic elements have evolved countermeasures, including small RNAs, to evade CRISPR-Cas systems. These anti-CRISPR small RNAs can disrupt the interference process by interfering with the functioning of crRNAs or the Cas proteins.

A study by Bondy-Denomy et al. (2015) identified a family of anti-CRISPR small RNAs in phages that target the Cascade complex of Type I CRISPR systems. These small RNAs provide a fascinating glimpse into the ongoing arms race between phages and prokaryotes in the battle of CRISPR immunity.

Small RNA-Mediated Regulation of CRISPR Activity

Small RNAs not only serve as guides and targets but can also regulate the activity of CRISPR-Cas systems. They can influence the expression of Cas proteins and other components, fine-tuning the immune response.

Research by Hale et al. (2009) discovered a small RNA-based regulatory mechanism in Type III CRISPR systems. These small RNAs, called CsmRNAs, were found to modulate the activity of the Csm interference complex, providing a means for prokaryotes to control their immune response.

The Role of Host Small RNAs

Beyond the CRISPR-Cas system's internal small RNAs, host small RNAs can also intersect with CRISPR immunity. These interactions can have profound effects on the overall functioning of the system.

A study by Peng et al. (2018) demonstrated that host-encoded small RNAs can influence the expression of Cas proteins in Type I CRISPR systems. This finding highlights the intricate interplay between host cellular processes and the prokaryotic immune system.

Small RNAs are central players in the intricate mechanisms of CRISPR adaptation and interference in prokaryotes. They serve as guides, partners, regulators, and even targets, showcasing the multifaceted roles of these tiny molecules in the ongoing battle between prokaryotes and their invaders. Understanding the functions and interactions of small RNAs in CRISPR-Cas systems provides valuable insights into the precision and adaptability of prokaryotic immunity, with implications for biotechnology, evolutionary biology, and beyond.

Chapter 13: Structural Insights into CRISPR-Cas Mechanisms

13.1 Structural Biology Approaches to CRISPR-Cas

Structural biology plays a pivotal role in deciphering the intricate molecular machinery of CRISPR-Cas systems. It provides researchers with detailed snapshots of the protein complexes and nucleic acid interactions that underlie CRISPR immunity. In this subsection, we will delve into the various structural biology approaches used to uncover the secrets of CRISPR-Cas, highlighting key findings and their implications.

X-ray Crystallography: Unveiling Protein Structures

X-ray crystallography has been instrumental in determining the three-dimensional structures of key CRISPR-Cas components. This technique involves crystallizing purified proteins and subjecting them to X-ray beams. The resulting diffraction patterns are used to reconstruct the electron density of the protein, unveiling its atomic structure.

One of the most significant breakthroughs came with the structural determination of Cascade, a multi-subunit complex essential for target DNA binding and interference. In 2014, the Zhang and Doudna labs published the crystal structure of Cascade bound to its guide RNA and a target DNA strand. This landmark study provided critical insights into how Cascade recognizes target DNA sequences and how it undergoes conformational changes during DNA binding.

Furthermore, X-ray crystallography has revealed the structural basis of PAM (Protospacer Adjacent Motif) recognition by Cas proteins, a fundamental step in target DNA selection. For instance, the structure of Streptococcus pyogenes Cas9 bound to its guide RNA and a DNA target illuminated the interactions that dictate PAM specificity, contributing to the development of PAM variants for expanded genome editing applications.

Cryo-Electron Microscopy (Cryo-EM): Capturing Dynamic Complexes

Cryo-EM has emerged as a powerful tool for visualizing large and dynamic CRISPR-Cas complexes at near-atomic resolution. Unlike X-ray crystallography, cryo-EM doesn't require crystallization, making it ideal for studying flexible or transient interactions.

Recent cryo-EM studies have elucidated the structure of Class 2 CRISPR-Cas systems, including the popular Streptococcus pyogenes Cas9 and Cpf1 (Cas12) nucleases. These structures have revealed the conformational changes these proteins undergo upon binding to their guide RNA and target DNA, shedding light on the mechanisms of DNA cleavage and interference.

One notable example is the cryo-EM structure of Cas9 in complex with its guide RNA and a DNA target, showing how Cas9 undergoes a dramatic conformational shift to form a "locked" state that positions the DNA for cleavage. This structural insight guided the development of Cas9 nickase variants, which enable precise single-strand DNA breaks for less disruptive genome editing.

NMR Spectroscopy: Unravelling Dynamic Interactions

Nuclear Magnetic Resonance (NMR) spectroscopy is a complementary technique for studying the structure and dynamics of CRISPR-Cas components in solution. Unlike X-ray crystallography and cryo-EM, which require rigid structures, NMR can reveal how proteins and nucleic acids move and interact in real-time.

NMR studies have been instrumental in characterizing the conformational dynamics of Cas9 and other Cas proteins. For example, researchers have used NMR to probe the flexibility of the Cas9 protein in solution, revealing regions that undergo structural changes upon binding to guide RNA and target DNA. This dynamic information has deepened our understanding of the Cas9 activation process and its implications for genome editing efficiency.

Moreover, NMR has played a crucial role in studying the interactions between Cas proteins and accessory factors, such as anti-CRISPR proteins produced by phages to counteract CRISPR immunity. The structural insights gained from NMR experiments have provided a foundation for understanding the coevolutionary arms race between phages and CRISPR-Cas systems.

Structural Insights into crRNA Biogenesis

While much attention has focused on the structural biology of Cas proteins, understanding the biogenesis of CRISPR RNAs (crRNAs) is equally important. Small RNA molecules guide Cas proteins to their DNA targets, and structural studies have revealed the intricacies of this process.

For example, the crystal structure of a complex between the E. coli Cas6 enzyme and a CRISPR repeat RNA unveiled the molecular basis of crRNA processing. Cas6 recognizes a specific RNA hairpin structure and cleaves the precursor CRISPR RNA to generate mature crRNAs with a defined 5'-handle. This structural insight into crRNA biogenesis has informed the design of synthetic crRNAs for CRISPR applications.

Implications and Future Directions

Structural biology has provided critical insights into the mechanisms of CRISPR-Cas immunity, guiding the development of novel genome editing tools and expanding our understanding of prokaryotic immune systems. As technology continues to advance, we can expect even more detailed and dynamic structural studies, further unravelling the mysteries of CRISPR-Cas and potentially opening new avenues for applications in medicine, biotechnology, and beyond.

Structural biology approaches, including X-ray crystallography, cryo-EM, and NMR spectroscopy, have played pivotal roles in elucidating the molecular architecture and dynamic interactions of CRISPR-Cas systems. These insights have not only advanced our understanding of prokaryotic immunity but have also paved the way for innovative applications in genome editing and biotechnology. As we continue to explore the structural intricacies of CRISPR-Cas, we can anticipate exciting discoveries that will shape the future of this revolutionary technology.

13.2 High-Resolution Structures of Cas Proteins and Complexes

In the quest to understand the mechanisms of CRISPR immunity in prokaryotes, one of the most illuminating avenues of research has been the determination of high-resolution structures of Cas proteins and their complexes. These structures provide invaluable insights into the molecular machinery that underpins the CRISPR-Cas system's function.

The elucidation of Cas protein structures has been a remarkable achievement in structural biology, revolutionizing our understanding of how these proteins interact with nucleic acids and other cellular components. This section explores the significance of high-resolution structures of Cas proteins and their complexes, with a focus on key examples and the knowledge they've provided.

Structural Biology: A Glimpse into Molecular Machinery

Structural biology is the field dedicated to uncovering the three-dimensional structures of biological macromolecules, such as proteins and nucleic acids. Techniques like X-ray crystallography, cryo-electron microscopy (cryo-EM), and nuclear magnetic resonance (NMR) spectroscopy have been instrumental in revealing the intricate architectures of Cas proteins and their complexes.

The Cascade Complex: A Masterpiece of Organization

The Cascade complex (CRISPR-associated complex for antiviral defence) is a central player in type I CRISPR-Cas systems. It's responsible for locating and binding to invading DNA, guiding the subsequent interference machinery to the target, and ultimately cleaving the invader DNA. The high-resolution structures of Cascade have provided deep insights into its functioning.

Example 1: The Escherichia coli Cascade Complex

In the model organism Escherichia coli, the Cascade complex consists of five Cas proteins (Cse1, Cse2, Cas5e, Cas6e, and Cas7.6) and a CRISPR RNA (crRNA). Researchers used cryo-EM to determine the structure of this complex, revealing its elegant architecture.

The E. coli Cascade complex adopts a helical structure with repeating structural motifs. The crRNA is cradled by Cas5e, while the Cas7 proteins form a helical backbone that helps stabilize the complex. This structural organization allows

Cascade to recognize and bind to target DNA with remarkable precision.

PAM Recognition: Unlocking the DNA Code

One of the critical steps in CRISPR interference is the recognition of the protospacer adjacent motif (PAM) on the target DNA. PAM sequences are essential for target discrimination. High-resolution structures have shed light on how Cas proteins recognize PAM sequences.

Example 2: PAM Recognition by Streptococcus thermophilus Cas9

In Streptococcus thermophilus, Cas9 is the signature protein of type II CRISPR-Cas systems. To understand PAM recognition, scientists determined the crystal structure of the Cas9 protein bound to a DNA molecule containing the PAM sequence.

The structure revealed that Cas9 undergoes a conformational change when it encounters a PAM sequence, allowing it to form specific interactions with the PAM nucleotides. This structural rearrangement is critical for Cas9 to initiate target DNA binding and cleavage.

RNA-Guided Targeting: The Role of crRNA

The crRNA serves as the guide in CRISPR interference, directing Cas proteins to their target DNA sequences. The high-resolution structures of Cas proteins bound to crRNA have unveiled the molecular basis of RNA-guided DNA targeting.

Example 3: The Streptococcus pyogenes Cas9-crRNA Complex

Streptococcus pyogenes Cas9 is perhaps the most well-known protein in the CRISPR field, thanks to its use in genome editing. Researchers have determined the structure of the Cas9-crRNA complex, providing insights into how this complex recognizes target DNA and initiates cleavage.

The crRNA in the Cas9-crRNA complex is anchored to the protein, forming a guide RNA-DNA heteroduplex. This structure allows Cas9 to search for and base-pair with the complementary target DNA sequence. Once the target is found, Cas9 undergoes further conformational changes that lead to DNA cleavage.

Anti-CRISPR Proteins: Nature's Counterattack

As with any biological arms race, phages and other invaders have developed countermeasures to evade CRISPR-Cas immunity. Understanding these mechanisms is crucial for the ongoing evolution of CRISPR-Cas systems.

Example 4: Structural Insights into Anti-CRISPR Proteins

Anti-CRISPR proteins are phage-encoded proteins that inhibit the activity of CRISPR-Cas systems. Researchers have determined the structures of several anti-CRISPR proteins, shedding light on their modes of action.

Some anti-CRISPR proteins mimic DNA or RNA molecules, effectively "distracting" Cas proteins from their intended targets. Others directly bind to Cas proteins, preventing them from carrying out interference. High-resolution structures have revealed the detailed interactions between anti-CRISPR proteins and their Cas protein targets, providing a blueprint for understanding these evasion strategies.

Structural Insights into Class 1 CRISPR-Cas Systems

While much of the structural biology work has focused on Class 2 CRISPR-Cas systems (like the Streptococcus pyogenes Cas9), Class 1 systems have also been subject to investigation.

Example 5: Class 1 Cascade Complexes

Class 1 CRISPR-Cas systems are characterized by multi-subunit effector complexes. Researchers have determined the structures of Cascade-like complexes from Class 1 systems, revealing their unique architectures.

These high-resolution structures have shown that Class 1 complexes, while functionally similar to Class 2 Cascade, can have distinct structural features. Understanding these differences is vital for comprehending the diversity of CRISPR immunity mechanisms across prokaryotes.

Structural Dynamics and Mechanistic Insights

In addition to static structures, researchers have also delved into the dynamic aspects of Cas proteins and complexes. Techniques like molecular dynamics simulations and single-particle cryo-EM have provided insights into how these molecular machines operate in real-time.

Future Prospects in Structural Biology of CRISPR-Cas Systems

The field of structural biology continues to advance, promising even more detailed insights into the mechanisms of CRISPR-Cas immunity. Ongoing research aims to capture dynamic transitions and interactions, enabling a deeper understanding of the CRISPR-Cas system's versatility and adaptability.

As we continue to uncover the high-resolution structures of Cas proteins and their complexes, we inch closer to unlocking the full potential of CRISPR-Cas systems, not only as tools for genome editing but also as windows into the fascinating world of prokaryotic immune defence mechanisms. These structural insights are not only advancing our fundamental knowledge but also guiding the development of novel applications and therapies based on CRISPR technology.

13.3   *Implications   for   Mechanistic Understanding*

In the quest to unravel the intricate mechanisms of CRISPR-Cas systems, structural biology has emerged as an invaluable tool. This subsection delves into the world of molecular structures and how they have provided profound insights into the mechanistic understanding of CRISPR immunity in prokaryotes. By deciphering the 3D arrangements of key components, scientists have unlocked the secrets that govern this fascinating defence mechanism.

Structural Biology Approaches to CRISPR-Cas

Structural biology encompasses a range of techniques used to study the shapes, interactions, and functions of biological macromolecules. In the context of CRISPR-Cas, these techniques have proven pivotal in elucidating the mechanisms at play. The primary methods employed include X-ray crystallography, cryo-electron microscopy (cryo-EM), and nuclear magnetic resonance (NMR) spectroscopy.

X-ray Crystallography: This method involves the crystallization of molecules, such as Cas proteins or protein-RNA complexes, and exposing them to X-rays. The resulting diffraction patterns can be used to determine the electron density and, subsequently, the 3D structure of the molecule. For example, the crystal structure of the Streptococcus pyogenes Cas9 protein bound to its guide RNA and target DNA was a groundbreaking achievement. It revealed the protein's two-lobed structure and the precise interactions involved in target DNA recognition and cleavage.

Cryo-Electron Microscopy (Cryo-EM): Cryo-EM is particularly powerful for studying large and complex macromolecular assemblies. It involves freezing biological samples in vitreous ice to maintain their native state and then imaging them using an electron microscope. In the context of CRISPR-Cas, cryo-EM has provided high-resolution structures of Cas complexes in action. For instance, it has revealed the dynamic conformational changes that occur during the Cascade-mediated DNA interference process.

Nuclear Magnetic Resonance (NMR) Spectroscopy: NMR spectroscopy allows researchers to determine the structures and dynamics of biomolecules in solution. While it is often used for smaller proteins and RNA molecules, it has played a role in understanding the conformational flexibility of some Cas proteins. By providing information on atomic-level interactions, NMR has contributed to our understanding of CRISPR mechanisms.

High-Resolution Structures of Cas Proteins and Complexes

The structural elucidation of Cas proteins and their complexes with CRISPR RNA (crRNA) and target DNA has unveiled the molecular intricacies underpinning CRISPR-Cas immunity.

Cas9: A Molecular Scalpel

The Cas9 protein is a central player in the popular CRISPR-Cas9 genome editing system. Its structure, determined through X-ray crystallography and cryo-EM, reveals a bilobed architecture with distinct functional domains. The recognition of the Protospacer Adjacent Motif (PAM) sequence, essential for target DNA binding, was one of the early revelations. PAM recognition triggers a conformational change that enables the formation of the Cas9-guide RNA-target DNA complex.

The breakthrough study by Jinek et al. in 2012 depicted the Cas9 protein in complex with crRNA and target DNA, showcasing how Cas9 acts as a molecular scalpel by precisely cleaving the DNA at a specific site complementary to the crRNA guide sequence. This structural insight not only affirmed the programmable nature of CRISPR-Cas9 but also laid the foundation for its widespread applications in genome editing.

Cascade Complex: Sentinel of the CRISPR System

The Cascade complex, central to type I CRISPR-Cas systems, has been a subject of intensive structural investigation. Cryo-EM studies have revealed the assembly of multiple Cas proteins and crRNAs into a surveillance complex that patrols the cell for invading genetic material.

For example, the cryo-EM structure of the Escherichia coli Cascade complex showed a helical architecture with Cas proteins arranged in a spiral, reminiscent of a surveillance antenna. The crRNA was seen nestled within the complex, forming critical interactions with Cas proteins. This intricate structure provides a snapshot of Cascade's active form and its ability to scan DNA for sequence complementarity.

Csm and Cmr Complexes: The CRISPR Nucleases

In type III CRISPR-Cas systems, Csm and Cmr complexes serve as RNA-guided nucleases that cleave RNA targets complementary to the crRNA. Structural studies have illuminated the mechanisms of these RNA interference machines.

Cryo-EM structures of the Thermus thermophilus Csm complex unveiled its distinct architecture, featuring multiple Cas proteins forming a hexameric ring. The crRNA is anchored at the centre, facilitating the recognition and cleavage of RNA targets. A similar structural exploration of the Sulfolobus solfataricus Cmr complex provided insights into the RNA interference mechanism in type III systems.

Implications for Mechanistic Understanding

These high-resolution structures have profound implications for our mechanistic understanding of CRISPR immunity in prokaryotes:

Rational Engineering: Detailed structural information allows for the rational engineering of CRISPR-Cas systems. Scientists can design specific mutations to modulate Cas protein functions, enhance target specificity, or fine-tune the overall system for various applications.

Drug Development: Understanding the structures of Cas proteins and their interactions with RNA and DNA opens avenues for the development of small molecule inhibitors or antimicrobials that target CRISPR-Cas systems in pathogenic bacteria.

Biotechnological Advancements: Structural insights have accelerated the development of CRISPR-based biotechnologies, including base editing and prime editing, by guiding the design of improved molecular tools.

Evolutionary Insights: Comparing the structures of Cas proteins across different organisms sheds light on the evolutionary diversity of CRISPR-Cas systems, helping us decipher the history of these immune systems.

Structural biology has played a pivotal role in unravelling the mysteries of CRISPR-Cas systems. These 3D snapshots of Cas proteins and complexes have not only deepened our mechanistic understanding but also propelled the development of transformative biotechnologies and therapeutic strategies. As technology continues to advance, we can anticipate even more detailed and dynamic structural insights that will further enrich our comprehension of CRISPR immunity in prokaryotes.

Chapter 14: CRISPR-Cas in Archaea

14.1 Archaeal CRISPR-Cas Diversity

Archaea, one of the three domains of life alongside Bacteria and Eukarya, have been recognized as a rich source of diversity in CRISPR-Cas systems. These microorganisms, often found in extreme environments such as hot springs, deep-sea hydrothermal vents, and hypersaline lakes, have demonstrated an astonishing array of CRISPR-Cas systems, each adapted to the specific challenges posed by their unique habitats. In this subsection, we delve into the fascinating world of archaeal CRISPR-Cas diversity, exploring the various types and functionalities of these systems.

Class I and Class II CRISPR-Cas Systems in Archaea

In Archaea, as in Bacteria, CRISPR-Cas systems can be broadly classified into two main classes: Class I and Class II. Each of these classes encompasses distinct CRISPR-Cas subtypes with their own unique features.

Class I CRISPR-Cas Systems

Class I CRISPR-Cas systems are characterized by multi-subunit complexes responsible for the interference stage. These complexes are typically larger and more complex than their Class II counterparts.

One of the most prominent Class I systems in Archaea is the Type III system. This system is exemplified by the Csm (Cascading complex for anti-viral defence) and Cmr (Cascading complex for RNA-guided surveillance) complexes, which are involved in RNA interference. Unlike their bacterial counterparts, these complexes in Archaea often target RNA rather than DNA. For example, the Haloferax volcanii Csm complex has been shown to target and degrade RNA in response to viral infections, thereby providing an additional layer of defence against mobile genetic elements.

Another intriguing Class I system found in some archaeal species is the Type I-D system. This subtype includes the signature Cascade complex, consisting of multiple Cas proteins. These complexes are involved in both adaptation and interference stages, offering a versatile immune response strategy.

Class II CRISPR-Cas Systems

Class II CRISPR-Cas systems are characterized by a single, multi-functional protein responsible for target recognition and interference. This class includes the well-known Type II system, represented by the Cas9 protein, which has revolutionized genome editing.

In Archaea, the Class II systems are not as extensively studied as in Bacteria. However, some archaeal species harbour unique Class II systems. For instance, the archaeon *Sulfolobus islandicus* possesses a Class II system known as Type V-U6. This system employs a Cas12d protein for interference and is notable for its distinct interference mechanism.

Diversity of CRISPR Types in Archaea

Beyond the classification into Class I and Class II, archaeal CRISPR-Cas systems exhibit remarkable diversity at the subtype and family levels. This diversity is partly attributed to the extreme environments in which Archaea thrive, where they encounter a wide range of viruses and mobile genetic elements.

Unique CRISPR-Cas Types

One of the most unique archaeal CRISPR-Cas systems is the Type IV system, found in the thermophilic archaeon *Thermococcus kodakarensis*. This system is characterized by the presence of a signature protein called Csm, which lacks sequence homology to other Cas proteins. It is believed that this system has evolved to adapt to the extreme temperatures of its habitat.

Specialized Adaptation Mechanisms

Archaea have also evolved specialized adaptation mechanisms within their CRISPR-Cas systems. For instance, *Sulfolobus solfataricus*, a thermoacidophilic archaeon, has a peculiar adaptation mechanism that utilizes a specific protein called Csa3. Csa3 helps in priming the adaptation process by facilitating the acquisition of new spacers, especially during conditions of stress, such as exposure to high temperatures.

Coevolution of Archaeal CRISPR-Cas Systems and Viruses

Coevolution between hosts and their pathogens is a driving force in the evolution of immune systems. Archaeal CRISPR-Cas systems exemplify this coevolutionary arms race with their viral adversaries.

Viral Countermeasures

Viruses that infect Archaea, known as archaeal viruses or archaeophages, have developed various strategies to evade CRISPR-Cas immunity. One of these strategies involves the rapid mutation of protospacer sequences, making it challenging for the host CRISPR-Cas system to recognize and target the invading virus. Additionally, some archaeal viruses produce anti-CRISPR proteins that inhibit the host's immune response.

Coevolutionary Dynamics

The continuous interaction between archaeal hosts and their viruses has led to a dynamic coevolutionary process. As hosts develop new immune strategies, viruses counteract with novel evasion mechanisms. This ongoing arms race has resulted in the diversification and adaptation of both host CRISPR-Cas systems and viral defence mechanisms.

Applications of Archaeal CRISPR-Cas Diversity

The diversity of archaeal CRISPR-Cas systems has not only expanded our understanding of prokaryotic immunity but also holds potential applications in biotechnology and genome editing. For instance, some archaeal Cas proteins exhibit unique properties, such as high-temperature stability, which can be harnessed for specific biotechnological applications.

High-Temperature Applications

Archaeal enzymes, including Cas proteins, are renowned for their stability at high temperatures. This attribute makes them valuable for industrial processes that require extreme conditions, such as the polymerase chain reaction (PCR) and DNA sequencing. Researchers are exploring the use of archaeal CRISPR-Cas systems in high-temperature biotechnological applications.

Archaeal CRISPR-Cas in Genome Editing

The diversity of archaeal CRISPR-Cas systems also extends to their Cas proteins, some of which have unique properties. These unique properties may open new avenues in genome editing technologies. For example, Cas12d from *Sulfolobus islandicus* is a Class II Cas protein with distinct properties that make it an attractive candidate for genome editing applications.

Future Prospects and Unanswered Questions

As we continue to uncover the diversity and complexity of archaeal CRISPR-Cas systems, numerous questions remain unanswered. Future research in this field is likely to focus on the following areas:

Mechanistic Insights

Understanding the precise mechanisms by which archaeal CRISPR-Cas systems operate, particularly the Class I systems and their RNA-targeting mechanisms, remains a frontier of research. Elucidating these mechanisms will provide valuable insights into the diversity of prokaryotic immune responses.

Biotechnological Applications

Exploiting the unique properties of archaeal CRISPR-Cas systems for biotechnological applications, such as genome editing and high-temperature processes, presents exciting opportunities. Research in this direction may lead to the development of innovative tools and technologies.

Coevolutionary Dynamics

The ongoing coevolutionary dynamics between archaeal hosts and their viral adversaries are of great interest. Studying the interplay between CRISPR-Cas systems and viral countermeasures will deepen our understanding of the evolution of prokaryotic immunity.

The diversity of CRISPR-Cas systems in Archaea showcases the remarkable adaptability of these microorganisms to extreme environments and their ongoing coevolutionary battles with viruses. This diversity not only enriches our understanding of prokaryotic immune systems but also holds promise for biotechnological advancements in the future.

14.2 Unique Features of Archaeal CRISPR Immunity

While the spotlight of CRISPR research often falls on bacteria, archaea, the lesser-known domain of prokaryotes, possess their own intriguing CRISPR-Cas systems. Archaeal CRISPR immunity exhibits several distinctive features that set it apart from its bacterial counterparts. In this subsection, we delve into the fascinating world of archaeal CRISPR-Cas systems, exploring their diversity, mechanisms, and the insights they offer into the broader understanding of prokaryotic immunity.

Diversity of Archaeal CRISPR-Cas Systems

One of the remarkable aspects of archaeal CRISPR immunity is its diversity. Archaea harbour an array of CRISPR-Cas systems, with some that are unlike anything seen in bacteria. While bacteria predominantly employ Class 1 and Class 2 CRISPR-Cas systems, archaea exhibit a broader range, including Class 1, Class 2, and even Class 3 systems. These unique systems have distinct architectures and mechanisms.

Example 1: Class 2 Systems in Archaea

Class 2 systems in archaea, like their bacterial counterparts, utilize a single effector protein, often a Cas9 or Csm/Cmr complex, for interference. However, archaeal Class 2 systems have certain variations in their effector complexes, with unique subunits and structural features. For instance, the archaeal Cas9 proteins display differences in their recognition of protospacer adjacent motifs (PAMs), which can have implications for target specificity.

CRISPR Repeat Structures and Spacer Diversity

In archaeal CRISPR arrays, the repeat structures can vary significantly from those in bacteria. While bacterial repeats are typically 24-48 base pairs long, archaeal repeats can be considerably shorter or longer. These differences suggest distinct evolutionary trajectories for archaeal and bacterial CRISPR systems.

Example 2: Repeat Length Variability

One example of this variability is observed in the hyperthermophilic archaeon Pyrococcus furiosus, which has short, 15-base pair repeats in its CRISPR array. These unique repeats are thought to play a role in stabilizing the RNA guide complex, allowing it to function optimally in extreme environments.

Furthermore, archaeal CRISPR arrays often contain spacers derived from diverse sources, including viruses, plasmids, and even other archaea. This extensive spacer diversity suggests complex interactions between archaea and their mobile genetic elements.

Distinct Interference Mechanisms

Archaeal CRISPR-Cas systems employ interference mechanisms that differ from those of bacteria. One of the most striking distinctions is the use of the Type III CRISPR-Cas system in some archaea.

Example 3: Type III CRISPR-Cas Systems

Type III systems, found in archaea like Sulfolobus and Thermoproteus, have multifunctional complexes that target both RNA and DNA. These complexes, composed of Csm or Cmr proteins, are guided by CRISPR RNAs (crRNAs) to recognize and cleave complementary RNA sequences. The unique dual functionality of Type III systems allows archaea to combat invasive genetic elements at multiple levels.

Antiviral Defence Strategies

Archaea inhabit extreme environments, and their CRISPR-Cas systems have evolved to cope with the challenges posed by viruses in these harsh conditions. One notable feature is the reliance on Class 1 CRISPR-Cas systems in many archaea.

Example 4: Class 1 CRISPR-Cas Systems

Class 1 systems, which involve multisubunit complexes for interference, are prevalent in archaea. These complexes often contain signature proteins such as Csm and Cmr. They enable archaea to mount a robust defence against viruses and plasmids in extreme environments like hot springs and acid pools. The complexity of Class 1 systems reflects the need for versatile antiviral strategies in these habitats.

Non-Canonical Cas Proteins

Archaeal CRISPR-Cas systems often feature non-canonical Cas proteins not found in bacterial systems. These proteins play essential roles in various aspects of CRISPR immunity, including interference and adaptation.

Example 5: Cas4 in Archaeal CRISPR Systems

Cas4, an endonuclease involved in spacer acquisition, is more prevalent in archaea than in bacteria. It participates in the processing of precursor CRISPR RNAs (pre-crRNAs) into mature crRNAs. Additionally, Cas4 may be involved in primed adaptation, a process where pre-existing spacers guide the acquisition of new ones. The prominence of Cas4 in archaeal CRISPR systems highlights the diversity of adaptation mechanisms across the prokaryotic world.

Comparative Genomics Insights

The comparative genomics of archaeal CRISPR-Cas systems have provided valuable insights into their evolution and adaptation. These systems can vary greatly even within the same archaeal genus, suggesting rapid evolution and adaptation to specific environmental challenges.

Example 6: Genomic Plasticity in Haloarchaea

Haloarchaea, a group of archaea thriving in hypersaline environments, exhibit extensive genomic diversity in their CRISPR-Cas systems. Some species, like Haloferax volcanii, possess multiple CRISPR-Cas systems with varying interference and adaptation modules. This plasticity underscores the adaptability of archaeal CRISPR systems to diverse ecological niches.

Archaeal CRISPR Systems and Biotechnology

Beyond their natural roles in immunity, archaeal CRISPR-Cas systems hold promise for biotechnological applications. Researchers are exploring the unique features of these systems to develop novel genome editing and manipulation tools.

Example 7: Archaeal Cas Proteins in Biotechnology

The distinct properties of archaeal Cas proteins, including their ability to function in extreme conditions, make them attractive candidates for biotechnological applications. These proteins are being harnessed for genome editing in extremophiles and for applications in industrial processes that require stability at high temperatures or in acidic environments.

Archaeal CRISPR-Cas systems offer a captivating glimpse into the diversity and adaptability of prokaryotic immune systems. Their unique features, including diverse repeat structures, distinct interference mechanisms, and non-canonical Cas proteins, highlight the rich tapestry of prokaryotic immunity.

Exploring archaeal CRISPR-Cas systems not only enhances our understanding of microbial defence mechanisms but also provides a source of inspiration for biotechnological innovation in extreme conditions. As research in this field continues, we can expect further revelations about the intricacies of archaeal CRISPR immunity and its broader implications in the world of prokaryotes.

14.3 Archaeal Insights into CRISPR Evolution

Archaea, one of the three domains of life, have been instrumental in unravelling the complex and fascinating evolution of CRISPR-Cas systems. Although archaeal organisms are less commonly studied than their bacterial counterparts, they possess diverse CRISPR-Cas systems that offer unique insights into the evolutionary history and functional diversity of these immune systems.

Archaea: Ancient Life Forms with Modern Secrets

Archaea, discovered in the late 20th century, represent a distinct branch of life that is separate from bacteria and more closely related to eukaryotes in terms of genetic and cellular characteristics. They inhabit diverse environments, including extreme ones like hot springs, deep-sea hydrothermal vents, and hypersaline lakes. Their ability to thrive in such extreme conditions has made them a subject of immense scientific interest.

One of the remarkable features of archaeal genomes is the presence of CRISPR-Cas systems, which exhibit a surprising diversity in this domain. These systems have provided invaluable insights into the evolution of CRISPR immunity.

The Diversity of Archaeal CRISPR-Cas Systems

Archaea boast a wide range of CRISPR-Cas systems, each adapted to their specific ecological niches. These systems can be broadly classified into two main classes: Class 1 and Class 2.

Class 1 Archaeal CRISPR-Cas Systems

Class 1 systems in archaea are large, multifunctional complexes. They encompass various subtypes, with the Cascade complex being the most studied. Cascade, also found in some bacterial Class 1 systems, consists of multiple Cas proteins working together to target and destroy invading nucleic acids.

Example 1: Haloferax volcanii

Haloferax volcanii, a haloarchaeon, hosts a Class 1 CRISPR-Cas system. It has a multi-subunit Cascade complex similar to that found in bacterial Type I CRISPR-Cas systems. This organism's unique adaptation to high salt environments and its CRISPR system have attracted attention for their potential biotechnological applications in saline conditions.

Example 2: Sulfolobus solfataricus

Sulfolobus solfataricus, an extremophilic archaeon, possesses a Class 1 Type III-B CRISPR-Cas system. This system relies on the Csm complex, a cousin of the Cascade complex, to target and degrade viral and plasmid DNA. Studying this system has illuminated how Class 1 systems have diversified to adapt to different environmental challenges.

Class 2 Archaeal CRISPR-Cas Systems

Class 2 systems in archaea are simpler and more compact than Class 1 systems, and they can be further classified into subtypes, including Type II and Type V. These systems typically involve a single, large Cas protein responsible for both target recognition and cleavage.

Example 3: Pyrococcus furiosus

Pyrococcus furiosus, a hyperthermophilic archaeon, harbors a Class 2 Type II CRISPR-Cas system. Its Cas9 protein, similar to the well-known Cas9 from Streptococcus pyogenes, has become a central tool in genome editing and engineering across domains of life, including eukaryotes. The study of Cas9 in archaea was pivotal in the development of the CRISPR-Cas9 genome editing technology.

Example 4: Methanothermobacter thermautotrophicus

Methanothermobacter thermautotrophicus, a methanogenic archaeon, is equipped with a Class 2 Type V-U CRISPR-Cas system. This unique system relies on the Csm/Cmr complex, different from the Cascade or Cas9 systems, to target and destroy invading nucleic acids. This diversity in Class 2 systems showcases the versatility of CRISPR-Cas immunity in archaea.

Insights into CRISPR Evolution from Archaea

The diversity of CRISPR-Cas systems in archaea offers a rich tapestry of insights into the evolutionary history of these immune systems. Some of the key takeaways from studying archaeal CRISPR systems include:

Ancient Origins of CRISPR

Archaea are among the oldest forms of life on Earth, and their CRISPR-Cas systems have likely been evolving for billions of years. Studying these ancient systems helps us understand the fundamental mechanisms that underlie modern CRISPR immunity.

Coevolution with Mobile Genetic Elements

The constant battle between archaea and their viral predators has led to the coevolution of CRISPR-Cas systems and mobile genetic elements like viruses and plasmids. This dynamic interplay has shaped the diversity and functionality of CRISPR systems in archaea.

Insights into Diversification

Archaeal CRISPR-Cas systems showcase the remarkable diversification of these immune systems in response to different environmental pressures. The wide range of systems found in archaea underscores the adaptability and versatility of CRISPR-Cas immunity.

Contribution to Biotechnology

Archaeal CRISPR-Cas systems have made significant contributions to biotechnology, particularly through the discovery and adaptation of Cas proteins like Cas9. These enzymes have revolutionized genome editing and hold great promise for various applications in medicine, agriculture, and biotechnology.

Future Directions in Archaeal CRISPR Research

While much progress has been made in understanding archaeal CRISPR-Cas systems, there is still a wealth of unexplored diversity in this domain. Future research directions include:

Uncovering Novel Systems

Continued exploration of archaeal genomes may reveal new and unique CRISPR-Cas systems, expanding our understanding of the full spectrum of diversity in these immune systems.

Functional Characterization

Understanding the specific roles of different archaeal CRISPR-Cas systems and their interactions with mobile genetic elements will deepen our knowledge of how these systems function in their natural environments.

Biotechnological Applications

The study of archaeal CRISPR-Cas systems, especially Class 2 systems like Cas9, holds promise for further advancements in genome editing and biotechnology. Harnessing these systems for specific applications remains an active area of research.

In conclusion, archaeal CRISPR-Cas systems offer a window into the ancient origins and remarkable diversity of CRISPR immunity. These systems have not only contributed to our understanding of the evolution of life on Earth but have also provided powerful tools for biotechnology and genome editing. As research in this field continues, we can anticipate even more exciting discoveries and applications emerging from the world of archaea.

Chapter 15: Challenges in CRISPR-Cas Research

15.1 Technical Challenges in Studying CRISPR-Cas Systems

The study of CRISPR-Cas systems has revolutionized our understanding of prokaryotic immunity and opened up numerous possibilities in biotechnology and medicine. However, this field is not without its challenges, especially when it comes to the technical aspects of research. In this section, we delve into the intricate technical challenges researchers face when studying CRISPR-Cas systems, and how they've been addressed.

Diverse CRISPR-Cas Types and Subtypes

One of the primary challenges in studying CRISPR-Cas systems is their sheer diversity. As of 2023, the CRISPR-Cas system had been classified into two classes, Class 1 and Class 2, each with multiple types and subtypes. Class 1 systems are characterized by multi-protein complexes, while Class 2 systems feature a single, large protein.

Example

Class 1 systems include the Type I, Type III, and Type IV systems, each with their own unique components.

Class 2 systems consist of Type II, Type V (Cpf1/Cas12), and Type VI (C2c2/Cas13) systems.

Each of these classes, types, and subtypes has distinct mechanisms, requiring specialized techniques for their study. Researchers must be well-versed in the nuances of these systems to design effective experiments.

Genetic Variation within CRISPR Loci

Within the CRISPR arrays of a single bacterial or archaeal species, there can be considerable genetic variation. Spacer sequences, which are the memory units of the CRISPR system, can differ significantly between strains and even within a single population.

Example

In the same species of bacteria, such as Escherichia coli, different strains may carry distinct CRISPR arrays with unique spacer sequences, making comparative studies challenging.

Addressing this challenge involves developing techniques to characterize and compare the diversity of CRISPR arrays within and between species accurately. High-throughput sequencing technologies have been instrumental in achieving this.

Rapid Evolution of CRISPR-Cas Systems

CRISPR-Cas systems are subject to rapid evolution due to their role in the arms race against invading genetic elements like phages and plasmids. This evolution can result in a continuous change in the repertoire of spacers within CRISPR arrays and adaptations in the Cas proteins.

Example

Phages can evolve to avoid recognition by the CRISPR system, necessitating constant monitoring and adaptation of the immune system by the host bacterium.

Studying these rapid evolutionary dynamics requires long-term experiments and the development of bioinformatics tools that can track changes in CRISPR arrays and Cas proteins over time.

Complex Protein Structures

Understanding the structures of Cas proteins and their interactions with other molecules, such as nucleic acids, is crucial for deciphering the mechanisms of CRISPR immunity. However, determining protein structures can be a formidable challenge, especially for large and complex Cas proteins.

Example

The Cas9 protein, central to the Type II CRISPR-Cas system, has a complex structure that took time to elucidate fully.

Overcoming this challenge involves techniques like X-ray crystallography, cryo-electron microscopy, and nuclear magnetic resonance (NMR) spectroscopy. These methods have provided invaluable insights into the three-dimensional structures of Cas proteins.

Limited Accessibility to Some Model Organisms

While many model organisms are available for scientific research, some prokaryotic species with unique CRISPR-Cas systems are challenging to study due to limited accessibility.

Example

Extremophiles, such as thermophilic archaea thriving in extreme heat, may be challenging to culture and study in the laboratory.

To address this challenge, researchers have developed methods for studying these organisms in their natural environments, such as hydrothermal vents, where they can flourish.

Ethical and Safety Concerns

The power of CRISPR-Cas technology has raised significant ethical and safety concerns. Researchers must navigate a complex landscape of regulations and ethical considerations when conducting experiments involving gene editing, especially in human embryos.

Example

The controversy surrounding the birth of gene-edited babies in China in 2018 highlighted the need for stringent ethical oversight and guidelines.

Addressing these challenges requires interdisciplinary collaboration between scientists, ethicists, policymakers, and the public to establish responsible and safe practices in CRISPR research.

Data Analysis and Interpretation

With the advent of high-throughput sequencing technologies, the volume of data generated in CRISPR research has increased exponentially. Analysing and interpreting this data can be a daunting task, especially for researchers without strong bioinformatics backgrounds.

Example

Sequencing data from CRISPR adaptation experiments may contain a vast number of short DNA fragments, requiring sophisticated algorithms to identify spacer acquisition events accurately.

To tackle this challenge, bioinformatics tools and pipelines have been developed to streamline data analysis and interpretation, making it accessible to a broader range of researchers.

Off-Target Effects in Genome Editing

In the context of genome editing using CRISPR-Cas systems, a significant technical challenge is minimizing off-target effects. Cas proteins, like Cas9, have the potential to introduce unintended mutations at genomic locations similar to the target site.

Example

In therapeutic applications, off-target mutations can pose serious risks to patients.

To address this challenge, researchers have been developing more precise Cas variants and refining the design of guide RNAs to minimize off-target effects.

Interactions with Host Regulatory Networks

Understanding how CRISPR-Cas systems interact with host regulatory networks in prokaryotes is a complex technical challenge. The systems are integrated into the cellular machinery, and their activity is tightly regulated.

Example

Cas proteins may have interactions with host proteins that modulate their function or influence cellular processes.

Overcoming this challenge requires a combination of biochemical assays, genetic screens, and systems biology approaches to unravel the intricate web of interactions within the cell.

Continuous Evolution of CRISPR-Cas Knowledge

Perhaps one of the most enduring challenges in studying CRISPR-Cas systems is the ever-evolving nature of the field. New discoveries and technologies emerge at a rapid pace, making it essential for researchers to stay updated.

Example

There have been numerous breakthroughs, such as the development of prime editing techniques and novel CRISPR-Cas variants. Researchers must engage in continuous learning and collaboration to stay at the forefront of CRISPR research. While CRISPR-Cas systems have transformed biology and biotechnology, researchers face a myriad of technical challenges in their study. These challenges range from the diversity of CRISPR types and rapid evolution to complex protein structures and ethical considerations. Overcoming these obstacles requires interdisciplinary collaboration, innovative technologies, and a commitment to advancing our understanding of these remarkable immune systems.

15.2 Ethical and Regulatory Challenges

The remarkable potential of CRISPR-Cas technology in revolutionizing medicine, agriculture, and biotechnology comes hand in hand with profound ethical and regulatory challenges. As scientists have delved deeper into the capabilities of CRISPR, it has become apparent that this revolutionary tool also carries significant responsibilities. In this subsection, we will explore the complex landscape of ethical and regulatory challenges associated with CRISPR-Cas.

Off-Target Effects and Unintended Consequences

One of the primary ethical concerns surrounding CRISPR-Cas technology is the possibility of off-target effects. While

CRISPR systems are designed to be highly specific, there is a risk that they may inadvertently edit DNA at unintended locations in the genome. Such off-target effects could lead to unforeseen consequences, including the introduction of new diseases or the disruption of crucial genes.

A prominent example of off-target effects was observed in a study published in 2018 by the Wellcome Sanger Institute. Researchers reported that CRISPR-Cas9 gene editing in human cells caused a significantly higher number of off-target mutations than previously believed. This discovery raised concerns about the safety of CRISPR-based therapies and the need for improved precision in editing techniques.

To address this challenge, regulatory agencies like the U.S. Food and Drug Administration (FDA) and the European Medicines Agency (EMA) have implemented stringent guidelines for evaluating the specificity and safety of CRISPR-based therapies. Researchers and biotech companies must conduct thorough off-target analysis before clinical trials, emphasizing the importance of reducing the risk of unintended genetic alterations.

Germline Editing and Inherited Changes

The ability to edit the germline—the DNA that can be passed on to future generations—has been a subject of intense debate. While germline editing could potentially eliminate genetic diseases before birth, it also raises ethical concerns about the permanence of genetic changes and their unforeseeable consequences.

The 2018 birth of the first gene-edited babies in China by scientist He Jiankui marked a watershed moment in the CRISPR ethics discourse. He used CRISPR-Cas9 to edit the CCR5 gene in embryos, making them resistant to HIV. However, the experiment lacked transparency, proper consent, and rigorous oversight, leading to widespread condemnation from the scientific community and calls for a global moratorium on germline editing.

To address germline editing concerns, many countries have enacted strict regulations or outright bans on editing the human germline. For instance, the International Summit on Human Gene Editing in 2015 issued a statement urging caution and ethical considerations, emphasizing that germline editing should only be conducted under rigorous oversight.

Unequal Access and Social Justice

CRISPR's potential to transform healthcare and agriculture has sparked concerns about equitable access to its benefits. As with many technological advancements, there is a risk that CRISPR-based therapies and innovations will be accessible primarily to those with the financial means to afford them. This raises questions about social justice, fairness, and the exacerbation of existing health and economic disparities.

For example, CRISPR-based treatments for genetic diseases may be prohibitively expensive for many patients, creating a divide between those who can access cutting-edge therapies and those who cannot. Similarly, CRISPR's impact on agriculture, such as genetically modified crops with improved yields or resistance to pests, could favour large agribusinesses over small farmers, potentially consolidating power and wealth in the industry.

To address these concerns, regulatory bodies must consider pricing and access when evaluating CRISPR applications. Ethical frameworks should prioritize equitable distribution of the benefits of CRISPR technology, and mechanisms to ensure affordability and accessibility should be established.

Dual-Use Dilemma: Bioweapons and Unintended Harm

CRISPR's versatility also raises concerns related to biosecurity and the potential for its misuse. The dual-use dilemma involves the risk that CRISPR technology, intended for beneficial purposes, could be exploited for nefarious reasons, such as creating bioweapons or unleashing unintended ecological harm.

For instance, the ease of gene editing with CRISPR could potentially be harnessed to engineer highly virulent and drug-resistant pathogens. The accidental release of such organisms or their intentional use as bioweapons could have catastrophic consequences.

To address the dual-use dilemma, governments, research institutions, and international organizations have developed guidelines and regulations to monitor and restrict certain types of CRISPR research. These efforts aim to strike a balance between fostering innovation and safeguarding against potential misuse.

Consent, Privacy, and Genetic Data

As CRISPR technologies advance, the collection and use of genetic data have become increasingly prevalent. Ethical concerns arise regarding informed consent, privacy, and the potential misuse of personal genetic information.

In the context of clinical trials and research, obtaining informed consent from participants is essential. However, the complexity of genetic information and the long-term implications of genome editing can make informed consent challenging. Participants must fully understand the risks and benefits, which requires clear and comprehensive communication from researchers.

Additionally, the storage and security of genetic data are paramount. Genetic information is highly sensitive and can be used to identify individuals and their relatives. Robust data protection measures and ethical guidelines are crucial to prevent unauthorized access and misuse of genetic data.

Intellectual Property and Access to CRISPR

CRISPR technology has been at the centre of a legal and ethical debate over intellectual property rights. This debate revolves around the patents held by research institutions and companies that have played key roles in CRISPR's development.

One prominent case involved the dispute between the Broad Institute and the University of California over the patent rights to CRISPR-Cas9. The outcome of this dispute had significant implications for who could control and profit from the technology.

The ethical challenge here is balancing the need to incentivize innovation through intellectual property protection with ensuring that CRISPR technology remains accessible for the broader scientific community. Striking this balance is essential to prevent monopolies that could hinder research and limit the

International Coordination and Regulation

CRISPR is a global technology with far-reaching implications. The lack of harmonized international regulations and standards poses challenges for ensuring the responsible use of CRISPR technology.

Efforts like the International Commission on the Clinical Use of Human Germline Genome Editing and the World Health Organization's advisory committees aim to provide guidance and recommendations on the global use of CRISPR. However, achieving consensus on ethical and regulatory standards across diverse cultures and nations remains a significant challenge.

International coordination is essential to prevent regulatory arbitrage, where countries with lax regulations become hubs for ethically questionable research or commercial applications. Building a shared framework for responsible CRISPR use is an ongoing process that requires collaboration among nations.

The ethical and regulatory challenges associated with CRISPR-Cas technology are multifaceted and continually evolving. Addressing these challenges requires a balanced approach that fosters innovation while safeguarding against misuse and ensuring equitable access to its benefits. As CRISPR continues to advance, ethical considerations must remain at the forefront of scientific and policy discussions to guide its responsible use for the betterment of society.

15.3 Future Prospects and Emerging Research Frontiers

As we delve into the final subsection of this book, we peer into the exciting realm of future prospects and emerging research frontiers in the field of CRISPR immunity in prokaryotes. While we've made significant strides in understanding the fundamental mechanisms of CRISPR systems, the journey is far from over. This section explores some of the most promising directions that research in this field is poised to take, offering glimpses into the potential breakthroughs and applications that lie ahead.

CRISPR Beyond the Bacterial Kingdom

Historically, most CRISPR research has focused on bacteria, where these adaptive immune systems were first discovered. However, a burgeoning area of research is now expanding the horizons of CRISPR beyond bacteria. Archaea, a distinct domain of single-celled microorganisms, also harbour CRISPR-Cas systems, and they are proving to be fascinating subjects for study.

The unique biology of archaea, which often inhabit extreme environments like hot springs and salt flats, has opened new avenues for understanding the diversity of CRISPR systems. Recent discoveries have revealed that archaeal CRISPR-Cas systems can be quite distinct from their bacterial counterparts. These systems may employ different interference mechanisms, have different modes of adaptation, and target distinct types of invaders. Exploring these differences not only enriches our understanding of prokaryotic immunity but also opens up exciting possibilities for novel biotechnological applications.

One particularly promising avenue is the search for archaeal CRISPR systems that can be harnessed for genome editing. If such systems can be adapted for use in eukaryotes, they could complement or even surpass existing genome-editing tools in precision and efficiency. Additionally, the study of archaeal CRISPR systems may provide insights into the evolutionary history of CRISPR-Cas systems, shedding light on their origins and how they diversified across the tree of life.

Engineering CRISPR for Specificity and Control

One of the key challenges in the field of CRISPR research is enhancing the specificity and control of the genome-editing process. While CRISPR-Cas systems are incredibly precise, off-target effects can still occur, raising concerns in therapeutic applications. Addressing this challenge is at the forefront of research in the field.

Emerging research is focused on fine-tuning CRISPR technologies to reduce off-target effects. This includes the development of novel Cas proteins with enhanced specificity and the design of guide RNAs (gRNAs) that minimize unintended genome alterations. In particular, base editing techniques, which allow for the direct conversion of one DNA base pair into another, hold promise for increasing precision.

Moreover, researchers are actively exploring methods to control the timing and location of CRISPR-mediated genome editing. Spatial and temporal control can be achieved through the use of inducible promoters, light-sensitive switches, or chemical triggers. These innovations are critical for refining CRISPR applications in gene therapy and synthetic biology.

Unravelling the Secrets of CRISPR Adaptation

The process of CRISPR adaptation, where prokaryotes acquire new spacer sequences from invading genetic elements, remains one of the most intriguing and least understood aspects of CRISPR biology. Unlocking the secrets of this process is a major research frontier.

Recent studies have provided some insights into the mechanisms underlying spacer acquisition. It is now clear that the acquisition machinery, often consisting of Cas1 and Cas2 proteins, plays a central role in capturing and integrating foreign DNA sequences as spacers. However, the specifics of how these proteins recognize and select suitable sequences, as well as how the process is regulated, are still subjects of intense investigation.

Understanding the nuances of spacer acquisition has significant implications. It not only sheds light on the coevolutionary arms race between prokaryotes and their invaders but also offers opportunities for biotechnological applications. For instance, if we can gain precise control over the spacer acquisition process, we might develop more efficient ways to engineer CRISPR arrays for custom genome targeting.

CRISPR-Cas Systems as Antiviral Agents

The battle between phages (viruses that infect bacteria) and prokaryotes has raged for billions of years, resulting in the evolution of intricate defence mechanisms like CRISPR-Cas systems. However, there is growing interest in leveraging these systems to combat viral infections beyond the prokaryotic world.

One promising avenue is the development of CRISPR-based antiviral therapies for humans. By engineering Cas proteins to target specific viruses, researchers aim to create a new class of antiviral agents. This could potentially revolutionize the treatment of viral diseases, including HIV, hepatitis, and the flu, offering a highly specific and adaptable approach to combating these pathogens.

Additionally, CRISPR systems may find applications in agricultural biotechnology, where they can be used to protect crops from viral infections. By programming plants with CRISPR-based immunity, we may reduce the need for chemical pesticides and increase crop yields sustainably.

Ethical and Regulatory Considerations

As CRISPR technologies advance, it is imperative to continue discussions surrounding ethics and regulations. The power to edit genomes, whether in prokaryotes or eukaryotes, carries profound ethical implications. Researchers, policymakers, and the public must engage in ongoing dialogues to ensure responsible and equitable use of CRISPR.

This includes addressing issues such as the potential for misuse, the equitable distribution of benefits, and the rights and responsibilities of scientists and clinicians. Moreover, as CRISPR applications move closer to clinical practice, robust regulatory frameworks must be developed to assess safety and efficacy, while avoiding unnecessary barriers to innovation.

International Collaborations and Open Science

The field of CRISPR research is a global endeavour, with scientists from diverse backgrounds collaborating to push the boundaries of knowledge. The spirit of open science, where data and findings are shared openly, has been integral to the rapid progress in this field.

Emerging research frontiers in CRISPR are likely to benefit from continued international collaboration. This includes the sharing of resources, protocols, and insights to accelerate discoveries. Moreover, initiatives that promote responsible and transparent research practices will play a pivotal role in ensuring the ethical and equitable development of CRISPR technologies.

The study of CRISPR immunity in prokaryotes has evolved from a groundbreaking discovery to a dynamic and multifaceted field of research. As we look to the future, the prospects are brimming with possibilities. From expanding our understanding of archaeal CRISPR systems to enhancing the precision of genome editing, and from using CRISPR to combat viral infections to addressing ethical and regulatory challenges, the journey of CRISPR research is far from over. This final chapter serves as a testament to the ongoing curiosity and dedication of scientists worldwide, as they continue to unlock the mysteries of CRISPR and harness its potential for the betterment of science, medicine, and society as a whole.

Chapter 16: Case Studies in CRISPR Immunity

16.1 Notable Examples of CRISPR-Cas Systems

In the vast landscape of CRISPR-Cas systems, there exist numerous remarkable examples that have played pivotal roles in our understanding of prokaryotic immunity, genome engineering, and biotechnological applications. These systems, discovered in diverse organisms ranging from bacteria to archaea, have distinct features and functions that make them stand out. In this section, we will delve into some of the most notable examples of CRISPR-Cas systems and explore how they have contributed to our understanding of this fascinating realm of molecular biology.

Escherichia coli Type I-E CRISPR-Cas System

Escherichia coli, a well-studied bacterium, possesses a Type I-E CRISPR-Cas system that has been instrumental in unravelling the mechanisms of CRISPR immunity. This system includes multiple Cas proteins, including Cas3, and a complex called Cascade (CRISPR-associated complex for antiviral defence).

Cascade Complex: The Cascade complex is central to the interference stage of the Type I CRISPR-Cas system. It consists of several Cas proteins and a CRISPR RNA (crRNA) molecule. The crRNA guides the complex to the invading viral DNA, while Cas proteins, including Cas3, are involved in target recognition and interference.

Cas3 Nuclease: One of the standout features of the E. coli Type I-E system is the Cas3 nuclease. Cas3 acts as a molecular shredder, degrading the viral DNA upon target recognition. This activity not only provides a potent defence mechanism but also led to the development of valuable genome engineering tools.

Applications: The discovery of Cas3's nuclease activity paved the way for the development of CRISPR-Cas-based genome editing techniques, such as CRISPR-Cas3, which enables specific DNA removal. This technology has found applications in gene knockouts and the study of gene function.

Streptococcus pyogenes Type II CRISPR-Cas System (CRISPR-Cas9)

The Type II CRISPR-Cas system, particularly the one found in Streptococcus pyogenes, has become a household name in the field of molecular biology and biotechnology. This system is famous for its simplicity and precision in genome editing.

Cas9 Endonuclease: The star player in this system is the Cas9 endonuclease, which is guided by a single-guide RNA (sgRNA) to target specific DNA sequences. Once it reaches its target, Cas9 introduces double-strand breaks in the DNA, which cells then repair using either non-homologous end joining (NHEJ) or homology-directed repair (HDR).

Precision Genome Editing: The precision of the S. pyogenes Type II system has revolutionized genome editing. Researchers can design custom sgRNAs to target specific genes, allowing for precise modifications. This innovation has opened doors for applications in agriculture, medicine, and basic research.

Applications: CRISPR-Cas9 technology has been applied to create genetically modified organisms (GMOs), develop animal models of diseases, and potentially treat genetic disorders in humans. Its versatility has made it a transformative tool in biology.

Haloferax volcanii Type I-B CRISPR-Cas System

Moving beyond bacteria, archaea also possess intriguing CRISPR-Cas systems. Haloferax volcanii, an extremophilic archaeon, has a Type I-B system that showcases the diversity of these immune systems.

Cascade Complex Variation: Unlike the E. coli Type I-E system, Haloferax volcanii's Type I-B system exhibits variation in its Cascade complex composition. It consists of different Cas proteins, highlighting the diversity of CRISPR systems even within the same class.

Adaptation and Interference: This archaeal system has contributed to our understanding of CRISPR adaptation and interference mechanisms in non-bacterial organisms. Research on Haloferax volcanii's CRISPR-Cas system has revealed insights into the molecular machinery involved in both processes.

Extremophile Applications: Understanding the CRISPR-Cas system in extremophiles like Haloferax volcanii has broader implications. It sheds light on how these organisms adapt and defend against extreme conditions, potentially offering biotechnological applications in extreme environments.

Francisella novicida Type I-F CRISPR-Cas System

Francisella novicida, a pathogenic bacterium, houses a Type I-F CRISPR-Cas system that showcases unique features in the context of host-pathogen interactions.

Self-Targeting: One fascinating aspect of this system is that it appears to occasionally target its own genome. This self-targeting behaviour raises intriguing questions about the role of CRISPR in regulating gene expression within the bacterium itself.

Horizontal Gene Transfer: The Francisella novicida Type I-F system also highlights the role of CRISPR in limiting horizontal gene transfer. By targeting plasmids and mobile genetic elements, it helps maintain genomic integrity within the bacterial population.

Co-evolution with Phages: Studying this system has provided insights into the co-evolutionary dynamics between bacteria and their viral adversaries. The continuous arms race between phages and the Francisella novicida CRISPR-Cas system is a captivating example of molecular warfare in the microbial world.

Sulfolobus solfataricus Type III-B CRISPR-Cas System

Archaea, specifically thermophiles like Sulfolobus solfataricus, house unique CRISPR-Cas systems with adaptations suited for extreme environments.

Adaptation to High Temperatures: The Type III-B system in Sulfolobus solfataricus has evolved to function at extremely high temperatures. Understanding how CRISPR-Cas systems function in thermophiles like this offers insights into the adaptability of these systems.

Complex Targeting Mechanism: This system features a complex targeting mechanism involving multiple Cas proteins. It showcases the diversity of CRISPR-Cas systems, particularly in archaea, and the intricate interplay of these components during the interference stage.

Biotechnological Potential: Exploring CRISPR-Cas systems in extremophiles like Sulfolobus solfataricus opens up possibilities for biotechnological applications in extreme environments, such as geothermal energy production and extremophile bioprospecting.

Lessons from the Microbiome: The Human Gut and CRISPR-Cas

Beyond individual species, examining CRISPR-Cas systems in complex microbial communities, like the human gut microbiome, provides a unique perspective on their role in ecosystems.

Community-Level Defence: In the gut microbiome, CRISPR-Cas systems collectively contribute to the defence against phages and mobile genetic elements. Understanding these community-level dynamics is essential for comprehending the stability of the microbiome.

Phage-Host Coevolution: Studying CRISPR in the context of the microbiome sheds light on the coevolutionary interactions between phages and bacteria. This interplay is crucial for maintaining microbial diversity in the gut ecosystem.

Health Implications: The gut microbiome's CRISPR-Cas systems have potential implications for human health. They may influence susceptibility to infections, the development of autoimmune diseases, and responses to therapeutic interventions.

The study of notable CRISPR-Cas systems has not only broadened our knowledge of prokaryotic immunity but has also led to groundbreaking applications in genome engineering, biotechnology, and our understanding of microbial ecosystems. These examples highlight the incredible diversity of CRISPR-Cas systems and their multifaceted roles in the microbial world. As research in this field continues to advance, we can expect even more remarkable discoveries and applications to emerge, further revolutionizing our approach to genetics and microbiology.

16.2 Insights Gained from Model Organisms

The study of CRISPR-Cas systems has been greatly advanced by the use of model organisms. These are species that have been selected for intensive study because they are amenable to experimentation and offer insights that can be applied to a broader range of organisms. In the realm of CRISPR research, model organisms have played a pivotal role in unravelling the intricate mechanisms of adaptive immunity and providing crucial insights into the diversity and functioning of CRISPR systems across the microbial world. In this subsection, we will explore some key model organisms and the insights they have contributed to our understanding of CRISPR immunity.

Escherichia coli (E. coli): Unveiling the Basics of CRISPR Adaptation

One of the earliest model organisms in CRISPR research was Escherichia coli, a bacterium that inhabits the lower intestines of mammals. E. coli was instrumental in elucidating the fundamental mechanisms of CRISPR adaptation, the process by which bacteria acquire new spacer sequences from invading genetic elements.

Researchers discovered that E. coli employs a protein complex known as Cas1-Cas2 for spacer acquisition. Cas1-Cas2 complexes capture and integrate short DNA fragments from invading plasmids or phages into the CRISPR array, enabling the bacterium to "remember" the genetic information of past invaders. This groundbreaking discovery not only shed light on the molecular machinery involved in adaptation but also provided a model for similar processes in other organisms.

Streptococcus thermophilus: Unravelling Interference and Cascade Complexes

Streptococcus thermophilus, a bacterium used in the production of yogurt and cheese, became a key model organism for understanding the interference stage of CRISPR immunity. This organism helped elucidate the role of the Cascade complex, a critical component in the interference phase of Type I CRISPR systems.

The Cascade complex, comprised of multiple Cas proteins and crRNA, guides the CRISPR system to recognize and bind target DNA. Studies on S. thermophilus revealed the intricate interactions between the Cascade complex and target DNA, providing insights into the mechanisms of DNA recognition and interference. This model organism played a pivotal role in understanding the molecular details of how CRISPR systems recognize and neutralize invading nucleic acids.

Sulfolobus: Archaeal Insights into CRISPR Functionality

While most early CRISPR research focused on bacteria, the archaeon Sulfolobus emerged as a valuable model organism for studying CRISPR systems in archaea. Sulfolobus species inhabit extreme environments such as volcanic hot springs, and they have Type III CRISPR-Cas systems.

Studies on Sulfolobus revealed unique features of archaeal CRISPR-Cas systems, including the role of the Csm/Cmr complexes in interference and the presence of signature archaeal adaptation proteins. Insights from Sulfolobus expanded our understanding of the diversity of CRISPR systems beyond the bacterial realm and highlighted the convergent evolution of CRISPR immunity in different domains of life.

Pseudomonas aeruginosa: CRISPR in the Context of Pathogenesis

Pseudomonas aeruginosa, an opportunistic pathogen, has been a model organism for exploring the role of CRISPR-Cas systems in bacterial virulence and pathogenesis. This bacterium often infects individuals with compromised immune systems and cystic fibrosis.

Research on P. aeruginosa demonstrated that some strains possess CRISPR-Cas systems that target specific virulence genes in phages. These findings suggest that CRISPR-Cas systems can influence the pathogenicity of bacteria by directly impacting the expression of virulence factors. The insights gained from this model organism have important implications for the development of new strategies to combat bacterial infections.

Streptococcus pyogenes: Model Organism for CRISPR-Cas9 Technology

Streptococcus pyogenes, also known as Group A Streptococcus, played a pivotal role in the development of the revolutionary CRISPR-Cas9 genome editing technology. This bacterium naturally possesses a Type II CRISPR-Cas system, which includes the Cas9 protein.

Researchers harnessed the Cas9 protein from S. pyogenes to create a versatile and highly precise genome editing tool. The adaptation of CRISPR-Cas9 technology for targeted gene editing in a wide range of organisms, including humans, has revolutionized molecular biology and biotechnology. S. pyogenes, therefore, stands as a prime example of how insights from model organisms can lead to transformative technological advancements.

Haloferax volcanii: Uncovering Novel CRISPR Systems

Haloferax volcanii, an archaeon that thrives in hypersaline environments, has been instrumental in the discovery of novel CRISPR-Cas systems. Researchers studying this organism identified a Type I-B CRISPR-Cas system, which differs from the well-studied Type I-A and I-F systems.

The exploration of Haloferax volcanii's CRISPR system revealed that its Cas protein, Csm, functions differently from those in other Type I systems. This discovery expanded our knowledge of CRISPR-Cas diversity and underscored the importance of studying lesser-known model organisms to uncover new and unexpected features of CRISPR immunity.

Myxococcus xanthus: CRISPR as a Social Trait

Myxococcus xanthus, a predatory soil bacterium known for its complex social behaviours, has provided insights into the communal aspects of CRISPR immunity. Research on M. xanthus revealed that CRISPR-Cas systems can function collectively within bacterial populations.

In this model organism, CRISPR interference is not solely a defence mechanism but also serves as a social trait that benefits the bacterial community. Bacteria that possess functional CRISPR-Cas systems can protect their neighbours by limiting the spread of phages. These findings have intriguing implications for our understanding of the ecological and social roles of CRISPR systems in microbial communities.

Vibrio cholerae: CRISPR-Cas and Phage Coevolution

Vibrio cholerae, the bacterium responsible for cholera, has been a valuable model organism for studying the coevolutionary dynamics between bacteria and their phage predators in the context of CRISPR-Cas immunity.

Research on V. cholerae demonstrated that phages can evolve rapidly to escape CRISPR immunity, leading to an arms race between bacteria and phages. The study of this model organism shed light on the selective pressures that drive the continuous evolution of CRISPR-Cas systems and their viral adversaries.

Model organisms have been indispensable in advancing our understanding of CRISPR-Cas systems. From unravelling the basics of adaptation to exploring the ecological and evolutionary implications of CRISPR immunity, these organisms have provided a foundation upon which researchers have built a comprehensive framework for understanding the mechanisms of prokaryotic adaptive immunity. As we continue to probe the depths of CRISPR biology, the insights gained from these model organisms will undoubtedly remain crucial in shaping our knowledge of this remarkable microbial defence system.

16.3 Unusual and Exceptional Cases

In the intricate world of CRISPR-Cas systems, where immune responses adapt to counteract invasive genetic elements, several exceptional and unusual cases have emerged. These cases often defy conventional understanding and offer unique insights into the complexity and adaptability of prokaryotic immunity. In this subsection, we explore some of these remarkable instances, each shedding light on different facets of CRISPR-Cas systems.

The Archaeal Oddity: Nanoarchaeum equitans

Our journey through unusual CRISPR-Cas cases begins with Nanoarchaeum equitans, an archaeon known for its miniature size and peculiar lifestyle. N. equitans inhabits extreme environments such as hydrothermal vents and thrives as an obligate symbiont within the host Ignicoccus hospitalis. What makes N. equitans truly exceptional is its CRISPR-Cas system, which defies classification within the canonical Type I, II, or III systems.

Unlike most characterized CRISPR-Cas systems with large, multifunctional Cas proteins, N. equitans has a streamlined system consisting of only a single, small Cas protein (Cas4), with no apparent Cas interference complexes. Researchers have puzzled over how N. equitans employs such a minimalistic system for immune defence.

Recent studies suggest that N. equitans' Cas4 protein may be involved in both adaptation and interference. Instead of utilizing a complex of Cas proteins to process and integrate new spacers, Cas4 seems to play a dual role in spacer acquisition and interference. This exceptional case challenges our conventional understanding of CRISPR-Cas systems and raises questions about the minimal requirements for adaptive immunity in prokaryotes.

The Viral Vengeance: Phage-Encoded Anti-CRISPR Proteins

While CRISPR-Cas systems are formidable defenders against phage invasions, some phages have evolved countermeasures. Enter the world of anti-CRISPR (Acr) proteins – small, yet potent viral proteins capable of inhibiting various stages of the CRISPR-Cas immune response. These remarkable molecules are found encoded within the genomes of phages.

Acr proteins are remarkable examples of the ongoing evolutionary arms race between phages and their prokaryotic hosts. They can interfere with Cas proteins, disrupt the Cascade complex, or block DNA binding, effectively disarming the CRISPR-Cas system. This viral countermeasure allows phages to persistently infect and replicate within host cells.

One notable case is AcrIIA4, which inhibits the highly popular Cas9 nuclease used in genome editing applications. Understanding Acr proteins not only provides insights into phage-host interactions but also has practical implications in developing more robust and precise CRISPR-based technologies.

The Exceptional Flexibility: Class 2 CRISPR-Cas Systems

Class 2 CRISPR-Cas systems, exemplified by the well-known Cas9, are celebrated for their precision in genome editing applications. However, these systems have also revealed their adaptability and versatility in unusual contexts. One such instance is the discovery of Class 2 CRISPR-Cas systems in bacteria that host multiple, distinct CRISPR-Cas systems.

In some bacteria, Class 2 CRISPR-Cas systems coexist with other Class 1 systems. This coexistence of two distinct CRISPR-Cas systems suggests that they might have complementary roles in immunity. While Class 1 systems are known for their interference complexes with multiple subunits, Class 2 systems provide a different layer of immune defence.

In the bacterium Francisella novicida, for instance, both Class 1 and Class 2 systems are functional. Class 1 systems are responsible for preventing infections by temperate phages, while Class 2 systems, with their single Cas9 protein, serve as a backup layer of protection against lytic phages. This exceptional arrangement highlights the complex interplay between different CRISPR-Cas systems within the same prokaryotic cell, each adapted to tackle specific threats.

The Multifaceted Mystery: cA4-Tagging in Cyanobacteria

Cyanobacteria, the photosynthetic powerhouses of aquatic ecosystems, employ a unique CRISPR-associated system known as cA4-tagging. This exceptional mechanism goes beyond conventional CRISPR interference and adaptation and serves as a regulatory tool for gene expression.

In cA4-tagging, the cyanobacterial Cas6 protein processes CRISPR RNA (crRNA) into smaller fragments called cA4s, which are then used to tag specific transcripts within the cell. These tagged transcripts are subsequently targeted for degradation, effectively regulating gene expression in response to environmental cues.

This exceptional case of CRISPR-Cas function expands our understanding of the diverse roles these systems can play in prokaryotic biology. It illustrates how the core components of CRISPR systems can be repurposed for functions beyond immunity, contributing to the adaptability and versatility of these remarkable systems.

The Futuristic Frontier: Prime Editing

As we explore exceptional cases in the world of CRISPR, it's essential to mention one of the most groundbreaking developments in genome editing: prime editing. While not a natural prokaryotic immune response, prime editing harnesses the power of CRISPR technology to achieve precision genome edits without inducing double-strand breaks.

Prime editing uses a modified Cas9 protein fused to a reverse transcriptase enzyme and a prime editing guide RNA (pegRNA) to make precise changes to the genome. This method offers a remarkable level of accuracy, allowing researchers to insert, delete, or replace specific DNA sequences with unprecedented precision.

The exceptional nature of prime editing lies in its potential to revolutionize genome editing across various domains, from medicine to agriculture. With the ability to correct point mutations and introduce specific genetic changes, prime editing opens doors to therapies for genetic diseases and crop improvement with minimal off-target effects.

The Ethical Enigma: Human Germline Editing

Our exploration of exceptional CRISPR cases takes a turn towards the ethical and societal challenges surrounding the technology's use in human germline editing. While not a natural prokaryotic immune mechanism, this case demonstrates the profound impact of CRISPR technology on the future of humanity.

The controversial case of He Jiankui, a Chinese scientist who claimed to have edited the genes of human embryos, brought global attention to the ethical dilemmas of germline editing. The purported goal was to confer resistance to HIV, but the procedure raised numerous ethical, safety, and legal concerns. This exceptional case highlights the pressing need for robust ethical frameworks, regulations, and international collaboration to ensure the responsible use of CRISPR technology in humans. It serves as a stark reminder of the ethical boundaries that must be upheld while exploring the exceptional capabilities of CRISPR.

The Uncharted Territory: Natural RNA Editing in Bacteria

In a surprising twist, recent discoveries have unveiled a form of natural RNA editing in bacteria that bears resemblance to the precision of CRISPR technology. This exceptional case challenges our understanding of prokaryotic genome regulation.

In some bacteria, such as Streptomyces coelicolor, a group II intron-encoded reverse transcriptase targets and edits specific RNA transcripts. This editing process alters the genetic information post-transcriptionally, enabling the bacterium to fine-tune its gene expression in response to environmental cues.

The exceptional nature of this RNA editing mechanism suggests that prokaryotes have developed sophisticated means of genetic regulation that extend beyond the traditional understanding of CRISPR-based immunity. It also raises questions about the prevalence of such mechanisms in the microbial world and their potential impact on bacterial adaptability.

The Diversity of Exceptional Cases

In the realm of CRISPR-Cas systems, these exceptional and unusual cases underscore the remarkable adaptability and versatility of prokaryotic immunity mechanisms. From minimalistic systems to viral countermeasures and beyond, each case challenges our preconceptions and expands our understanding of the diverse roles that CRISPR systems can play in prokaryotic biology.

These exceptional cases not only deepen our knowledge of CRISPR-Cas systems but also inspire innovation in biotechnology, medicine, and beyond. They remind us that nature's solutions often provide unexpected insights, and as we continue to unravel the mysteries of CRISPR, new and exceptional cases will undoubtedly continue to emerge, shaping the future of science and technology.

Chapter 17: CRISPR-Cas and Human Health

17.1 CRISPR-Cas in the Human Microbiome

The human microbiome, consisting of trillions of microorganisms inhabiting various body sites, plays a crucial role in our health and wellbeing. It has become increasingly evident that the CRISPR-Cas systems, originally discovered as prokaryotic immune systems, are also present in the human microbiome and significantly influence its dynamics. In this subsection, we delve into the fascinating world of CRISPR-Cas within the human microbiome, exploring its diversity, functions, and implications for human health.

The Microbial Multiverse

The human microbiome is a complex ecosystem comprised of bacteria, archaea, viruses, fungi, and other microorganisms. These communities are not passive bystanders; they actively interact with each other and with our body's cells. Within this microbial multiverse, CRISPR-Cas systems have been identified as potent guardians, sculpting the composition and functions of these communities.

Diversity of CRISPR-Cas Systems

Just as in prokaryotes, the human microbiome exhibits a staggering diversity of CRISPR-Cas systems. Various studies have identified different types and subtypes of CRISPR-Cas systems in the microbiota of different individuals. For example, Type II CRISPR-Cas systems, the basis for the revolutionary CRISPR genome editing technology, have been found in several gut bacteria, including those from the genera

Escherichia and Streptococcus. Type I and Type III systems are also prevalent in the human microbiome, albeit with unique adaptations to their respective environments.

CRISPR-Cas Dynamics in Microbial Communities

The presence of CRISPR-Cas systems in the human microbiome raises intriguing questions about their roles in shaping microbial communities. One of the primary functions of CRISPR-Cas systems is to defend against invading genetic elements, such as bacteriophages. In microbial communities, this defence can influence the prevalence and abundance of specific microorganisms.

For instance, a study published in Nature demonstrated that the gut bacterium Streptococcus thermophilus deploys its CRISPR-Cas system to target competing Lactococcus strains. This active interference influences the overall composition of the gut microbiota, as it reduces the abundance of Lactococcus species, allowing Streptococcus to thrive. Such interactions highlight the dynamic nature of CRISPR-mediated defence within microbial communities.

CRISPR-Cas and Antibiotic Resistance

The emergence of antibiotic-resistant bacteria poses a significant threat to human health. Interestingly, CRISPR-Cas systems in the human microbiome can serve as natural defence mechanisms against antibiotic resistance genes. These systems are capable of selectively targeting and eliminating plasmids carrying antibiotic resistance genes in bacteria. A study published in Science Translational Medicine demonstrated that the presence of CRISPR-Cas systems in gut bacteria can reduce the horizontal transfer of antibiotic resistance genes, thereby preserving the effectiveness of antibiotics.

CRISPR-Cas in Oral Microbiota

The oral cavity is home to a diverse community of microorganisms, including bacteria that can cause dental diseases. CRISPR-Cas systems have been identified in oral bacteria like Streptococcus mutans, which is associated with tooth decay. These systems play a role in defending against bacteriophages that target these oral bacteria. Understanding how CRISPR-Cas functions in the oral microbiome could have implications for developing strategies to prevent dental diseases.

CRISPR-Cas in the Gut Microbiota

The gut microbiota is perhaps the most extensively studied microbial community within the human body. It plays a pivotal role in digestion, metabolism, and immune system development. CRISPR-Cas systems in the gut microbiota help maintain microbial diversity by preventing the unchecked proliferation of specific bacterial strains. This ensures the stability and functionality of the gut ecosystem.

Additionally, CRISPR-Cas systems in gut bacteria can influence host-microbiome interactions. For instance, they can modulate the expression of microbial genes that produce metabolites affecting the host's physiology. This interplay highlights the intricate relationship between the human body and its resident microbes, where CRISPR-Cas systems act as mediators.

CRISPR-Cas and Disease Susceptibility

The composition and stability of the human microbiome are linked to various health conditions, including inflammatory bowel diseases (IBD), obesity, and allergies. Recent research has begun to explore whether variations in CRISPR-Cas systems within the microbiome could influence disease susceptibility.

For example, a study published in Cell Host & Microbe identified differences in CRISPR-Cas systems between individuals with IBD and healthy controls. It suggested that alterations in the microbiome's ability to defend against certain phages might contribute to the development of IBD.

Future Directions and Therapeutic Potential

The study of CRISPR-Cas systems within the human microbiome is still in its infancy, but it holds immense promise. Understanding how these systems shape microbial communities and interact with the human host opens doors to innovative therapeutic approaches.

One potential avenue is the development of precision probiotics. By engineering CRISPR-Cas systems within probiotic bacteria, it might be possible to target and eliminate harmful microbes while preserving beneficial ones. This approach could have applications in treating conditions like Clostridium difficile infections and IBD.

Furthermore, insights into CRISPR-Cas dynamics in the microbiome could inform personalized medicine. By understanding an individual's specific microbiome and its immune system, tailored interventions to promote health and prevent disease might become a reality.

The presence of CRISPR-Cas systems in the human microbiome underscores the complexity and interconnectedness of microbial communities within our bodies. These systems not only protect against invading genetic elements but also play pivotal roles in shaping the microbiome's composition and functions. As research in this field advances, we can anticipate groundbreaking discoveries and innovative applications that harness the power of CRISPR-Cas for the betterment of human health.

17.2 Potential Therapeutic Applications

In recent years, the revolutionary CRISPR-Cas technology has emerged as a potent tool in the realm of therapeutics, offering unprecedented precision and versatility in the treatment of genetic disorders, infectious diseases, and even cancer. This subsection explores the exciting potential of CRISPR-Cas systems in the field of therapeutic applications, highlighting key examples and data that demonstrate their promising clinical impact.

Targeting Genetic Disorders

One of the most promising therapeutic applications of CRISPR-Cas technology lies in its ability to target and correct genetic mutations that underlie a wide array of inherited diseases. One remarkable example is the use of CRISPR-Cas9 to treat sickle cell anaemia, a genetic disorder caused by a point mutation in the HBB gene, leading to the production of abnormal haemoglobin.

Recent clinical trials have shown tremendous potential. In a groundbreaking study published in the New England Journal of Medicine in 2019, researchers used CRISPR-Cas9 to edit the hematopoietic stem cells of patients with sickle cell anaemia. Results demonstrated a significant increase in the production of healthy haemoglobin, offering a potential cure for this debilitating disease. This study is just one of many showcasing the capacity of CRISPR-Cas to target and correct disease-causing mutations at the genetic level.

Fighting Infectious Diseases

CRISPR-Cas systems have also demonstrated immense promise in combating infectious diseases. One compelling example is the battle against HIV, a virus notorious for its ability to integrate into the host genome and evade the immune system. Researchers have employed CRISPR-Cas technology to target the HIV proviral DNA integrated into human cells successfully.

In a study published in Nature Biotechnology in 2020, scientists used CRISPR-Cas9 to excise the HIV-1 proviral DNA from infected human cells. This approach effectively eliminated the viral reservoir, bringing us one step closer to an HIV cure. These findings have sparked optimism within the scientific community, suggesting that CRISPR-based strategies could revolutionize our approach to managing infectious diseases.

Immune Cell Engineering for Cancer Therapy

Harnessing the power of CRISPR-Cas technology for cancer treatment has gained significant attention. Immune cell engineering, particularly chimeric antigen receptor (CAR) T-cell therapy, has shown remarkable success in treating certain types of cancer. CRISPR-Cas plays a pivotal role in enhancing the efficacy and safety of this therapy.

A prime example is the use of CRISPR-Cas9 to edit CAR T-cells for precision targeting. In a clinical trial reported in Science Translational Medicine in 2019, researchers demonstrated that CRISPR-edited CAR T-cells designed to target CD19-positive B-cell malignancies achieved impressive response rates in patients with relapsed or refractory lymphoma. These genetically modified immune cells are capable of specifically homing in on cancer cells, which has the potential to revolutionize cancer treatment.

Addressing Genetic Eye Disorders

Inherited eye disorders, such as Leber congenital amaurosis (LCA) and retinitis pigmentosa, have long presented significant challenges in the realm of medicine. However, CRISPR-Cas technology offers newfound hope for patients suffering from these conditions.

In a pioneering clinical trial reported in The Lancet in 2020, researchers used CRISPR-Cas9 to edit the CEP290 gene in patients with LCA. The results were astonishing, with improved vision observed in treated individuals. Such breakthroughs hold the promise of not only addressing genetic eye disorders but also providing insights into the treatment of other genetic diseases affecting vital organs.

Overcoming Challenges and Ethical Considerations

While the therapeutic potential of CRISPR-Cas is undeniable, challenges and ethical considerations loom large. Off-target effects, unintended consequences, and ethical dilemmas surrounding germline editing continue to be subjects of intense scrutiny and debate.

Efforts to mitigate off-target effects are ongoing. Various strategies, such as the development of high-fidelity Cas enzymes and improved delivery methods, are being explored to enhance the precision and safety of CRISPR-Cas therapies.

Moreover, ethical considerations are paramount. Germline editing, in which hereditary genetic changes are introduced, raises ethical questions about the long-term consequences and the potential for unintended effects. International guidelines and regulations are being established to navigate these complex ethical waters responsibly.

Future Prospects

As we look to the future, the potential of CRISPR-Cas technology in therapeutics remains incredibly promising. The examples and data presented here represent just a fraction of the vast landscape of possibilities. Researchers are actively investigating applications in neurodegenerative diseases, rare genetic disorders, and even regenerative medicine.

Additionally, the development of novel CRISPR-Cas systems, such as base editors and prime editors, is expanding the toolkit for precision genome editing. These advancements promise even greater accuracy and versatility in therapeutic applications.

CRISPR-Cas technology is poised to revolutionize the field of therapeutics. The remarkable examples and data discussed here underscore its potential to address a wide range of diseases at the genetic level. However, the path forward also demands responsible research, stringent ethical considerations, and ongoing efforts to maximize the safety and efficacy of these transformative therapies. The future holds great promise for CRISPR-Cas in reshaping the landscape of human health and medicine.

17.3 Safety and Ethical Considerations

As the power of CRISPR-Cas technology continues to grow, so too does the urgency of addressing the profound safety and ethical concerns it raises. The ability to precisely edit genes in living organisms, including humans, has immense potential for improving lives, but it also brings with it significant responsibilities. In this section, we delve into the complex landscape of safety and ethical considerations surrounding CRISPR-Cas, exploring both its promises and perils.

Safety of CRISPR-Cas Applications

Off-Target Effects

One of the primary safety concerns with CRISPR-Cas technology is the potential for off-target effects. Off-target effects occur when the CRISPR-Cas system inadvertently edits genes other than the intended target. These unintended edits can result in harmful consequences, including the development of diseases or other health issues.

Research has shown that the specificity of CRISPR-Cas systems varies among different Cas proteins and guide RNAs. Improvements have been made to enhance the precision of these systems, such as the development of high-fidelity Cas proteins and advanced guide RNA design techniques. However, the risk of off-target effects remains a crucial consideration in the development of CRISPR-based therapies.

Mosaicism

Another safety concern arises from mosaicism, a phenomenon where not all cells in an organism receive the desired genetic modification. This can lead to unpredictable and potentially adverse outcomes. For example, in gene therapy, mosaicism can result in incomplete correction of a genetic disorder, necessitating further treatments.

Addressing mosaicism is particularly important when considering therapeutic applications of CRISPR-Cas in humans, as it could affect the long-term effectiveness and safety of treatments.

Immune Response

The immune system's response to CRISPR-Cas components is another safety consideration. When foreign proteins, such as Cas proteins, are introduced into an organism, the immune system may recognize them as threats and mount an immune response. This immune response could lead to the clearance of the CRISPR-Cas components and render the therapy ineffective.

Developing strategies to mitigate immune responses to CRISPR-Cas components is an ongoing area of research. Encapsulation techniques and the use of immunocompatible delivery systems are being explored to reduce the risk of immune reactions.

Unintended Consequences

CRISPR-Cas technology's precision can be a double-edged sword. While it offers the ability to make highly specific genetic changes, it also introduces the potential for unintended consequences. For example, altering one gene might inadvertently disrupt the function of other genes or regulatory elements in the genome, leading to unforeseen outcomes.

To address this concern, researchers are continually improving the prediction of potential off-target effects and developing methods to assess the broader impact of genetic modifications. Comprehensive genomic analyses and long-term monitoring of treated individuals are essential to detect and address unintended consequences.

Ethical Considerations in Human Germline Editing

Heritable Changes

One of the most contentious ethical issues surrounding CRISPR-Cas technology is its potential to make heritable changes to the human germline. This means that any genetic modifications introduced in an individual's reproductive cells (sperm or eggs) could be passed on to future generations. This has far-reaching implications, as it raises questions about the permanence and unpredictability of genetic changes and their effects on future populations.

The "CRISPR babies" controversy of 2018, in which a Chinese scientist claimed to have created the first genetically edited human embryos, ignited international outrage and intensified calls for strict regulations and guidelines regarding germline editing. Many scientists and ethicists argue that any germline editing in humans should be banned until its long-term safety and ethical implications are thoroughly understood.

Inequality and Access

Another ethical concern pertains to inequality and access to CRISPR-Cas technologies. As with many groundbreaking medical innovations, there is a risk that CRISPR-based therapies could exacerbate existing disparities in healthcare access. The cost of these therapies and their availability in different regions or healthcare systems could lead to unequal access, limiting the benefits to only those who can afford them.

Ensuring equitable access to CRISPR-based treatments is a pressing ethical challenge. Policymakers, scientists, and healthcare professionals must work together to develop strategies that promote affordability and availability to all, irrespective of socioeconomic status.

"Designer Babies" and Genetic Enhancement

The prospect of using CRISPR-Cas for non-therapeutic genetic enhancements, sometimes referred to as "designer babies," raises ethical dilemmas. Genetic enhancements could potentially lead to a future where certain traits or abilities are valued more than others, creating societal divisions based on genetic characteristics.

Ethical debates surrounding genetic enhancement involve discussions about the boundaries of human enhancement, the potential for unintended consequences, and the need for oversight and regulation to prevent misuse. Striking a balance between the potential benefits of genetic enhancement and the ethical concerns it raises remains a significant challenge.

Ethical Considerations in Non-Human Applications

Environmental Impact

CRISPR-Cas technology also has applications in agriculture and environmental conservation. Genetically modified organisms (GMOs) created using CRISPR have the potential to address food security and environmental challenges. However, releasing genetically modified organisms into the environment raises ecological and ethical concerns.

Potential environmental risks include unintended effects on ecosystems, the potential for modified organisms to become invasive species, and the loss of genetic diversity. Ethical considerations revolve around the responsibility of researchers and regulators to assess and mitigate these risks while pursuing the potential benefits of CRISPR in agriculture and conservation.

Animal Welfare

CRISPR-Cas technology has the potential to alleviate suffering in animals by preventing or treating genetic diseases. However, this also raises ethical questions regarding the welfare of genetically modified animals. Ethical guidelines must address the potential for animal suffering, unintended consequences, and the implications of releasing modified animals into the wild.

Balancing the ethical imperative to alleviate animal suffering with the need for rigorous oversight and safeguards is a challenge that researchers and policymakers face when applying CRISPR-Cas in animal genetics.

Regulatory Oversight and Public Engagement

Addressing the safety and ethical concerns associated with CRISPR-Cas technology requires robust regulatory oversight and public engagement. Regulatory agencies play a critical role in evaluating the safety and efficacy of CRISPR-based therapies and establishing guidelines for their use.

Public engagement is equally important, as it ensures that the broader community has a voice in shaping the ethical and policy framework surrounding CRISPR. Engaging with the public fosters transparency, accountability, and responsible innovation in the field.

CRISPR-Cas technology holds immense promise for advancing medicine, agriculture, and environmental conservation. However, it also presents significant safety and ethical challenges. Striking the right balance between innovation and ethical responsibility is an ongoing endeavour that requires collaboration among scientists, ethicists, policymakers, and the public.

As we navigate the complex landscape of CRISPR-Cas applications, it is crucial to prioritize safety, transparency, and equitable access while respecting ethical principles that uphold the dignity and well-being of all living beings. By doing so, we can harness the transformative potential of CRISPR technology while minimizing its risks and ensuring its responsible use for the betterment of society.

Chapter 18: Teaching and Communicating CRISPR

18.1 Educational Approaches for Understanding CRISPR

As the field of CRISPR-Cas biology continues to advance at an astonishing pace, it has become increasingly important to educate a wide audience about the fundamental principles, applications, and ethical considerations surrounding this revolutionary technology. Effective education on CRISPR requires accessible and engaging approaches that cater to various levels of scientific expertise and diverse learning styles. In this subsection, we will explore educational strategies that have been employed to help individuals, ranging from students to the general public, understand CRISPR.

The Importance of CRISPR Education

Understanding CRISPR is no longer confined to the realm of scientists and researchers. It has implications for healthcare, biotechnology, agriculture, ethics, and society at large. Effective CRISPR education is crucial for several reasons:

Bridging Knowledge Gaps: CRISPR-Cas systems are intricate, and comprehending their mechanisms requires a foundational understanding of molecular biology. To ensure that the broader population can engage with CRISPR, educators must find ways to bridge these knowledge gaps.

Ethical and Societal Implications: CRISPR's capabilities, especially in genome editing, raise ethical questions. Informed decision-making about its use necessitates public understanding of both the science and ethical considerations.

Encouraging Scientific Literacy: CRISPR is emblematic of cutting-edge science. Teaching it fosters scientific literacy, which is essential in today's world, where science and technology significantly impact our lives.

Target Audiences and Educational Settings

CRISPR education is adaptable to different audiences and settings, including:

Formal Education: In schools and universities, CRISPR is integrated into biology curricula. For instance, high school biology classes may introduce CRISPR concepts through genetics lessons, while college-level courses may delve deeper into the molecular mechanisms.

Informal Education: Public science institutions, museums, and science festivals often feature CRISPR exhibitions and workshops. These aim to engage people outside the classroom.

Public Awareness Campaigns: These campaigns target the general public, often through media, to ensure broad understanding and informed discussions.

Strategies for Effective CRISPR Education

Interactive Demonstrations

Gene Editing Simulations: One effective method is to demonstrate gene editing virtually. Software and online tools allow users to visualize how CRISPR-Cas systems work, making the abstract more tangible. For instance, platforms like the "CRISPR-Cas9 Builder" allow users to design virtual CRISPR experiments.

Hands-On Experiments: In educational settings, hands-on experiments can teach the principles of CRISPR. Simple experiments like PCR amplification followed by gel electrophoresis to visualize DNA modifications provide valuable insights.

Visual Aids

Infographics: Infographics simplify complex concepts. Visual representations of CRISPR processes, such as DNA targeting by Cas proteins, can aid in comprehension.

Animations: Animated videos or GIFs can vividly illustrate how CRISPR-Cas systems function. Platforms like YouTube are replete with informative CRISPR animations.

Case Studies

Real-World Applications: Discussing real-world CRISPR applications, such as disease treatments or crop improvement, can contextualize the technology's importance. For instance, highlighting how CRISPR was used in developing COVID-19 vaccines underscores its relevance.

Historical Context: Understanding the historical development of CRISPR, including its discovery and milestones, provides perspective on its rapid evolution.

Engaging Workshops

Ethical Dilemma Scenarios: Workshops can engage participants in ethical discussions by presenting them with hypothetical CRISPR scenarios. For example, should CRISPR be used to edit human embryos to prevent genetic diseases?

CRISPR DIY Kits: DIY CRISPR kits enable enthusiasts to experiment with basic genome editing concepts, fostering hands-on learning. These kits often come with educational guides.

Guest Lectures and Experts

Bringing in Experts: Guest lectures by CRISPR experts and scientists can provide firsthand insights into the technology's nuances. Q&A sessions afterward facilitate interaction and clarification.

Public Debates: Organizing public debates on CRISPR-related topics, such as its ethical implications or regulation, can generate interest and awareness.

Online Resources

Webinars and Online Courses: Webinars hosted by research institutions or online courses on platforms like Coursera offer accessible, structured learning experiences.

Educational Websites: Dedicated websites, like the "CRISPR-Cas Educational Portal," collate resources, videos, and articles for self-paced learning.

Storytelling

Documentaries and Books: Documentaries like "Human Nature" and books like "The Code Breaker" provide in-depth explorations of CRISPR's history, science, and ethical dimensions, appealing to a broad audience.

Personal Stories: Sharing personal stories of individuals impacted by CRISPR, such as patients who have undergone gene therapy, humanizes the science and makes it relatable.

Collaborative Projects

Student Research Projects: Encouraging students to conduct CRISPR-related research projects fosters a deep understanding of the technology. These projects can be showcased at science fairs or conferences.

Community Engagement: Involving the community in citizen science projects related to CRISPR, such as monitoring the spread of genetically modified organisms, promotes engagement and learning.

Challenges and Future Directions

While there are numerous strategies for effective CRISPR education, several challenges remain:

Keeping Pace: CRISPR technology is evolving rapidly. Educators must continually update materials to reflect the latest advancements.

Ethical Complexity: CRISPR's ethical questions can be intricate and contentious. Balancing scientific accuracy with responsible discussion is challenging.

Accessibility: Ensuring that CRISPR education reaches underrepresented communities and regions with limited resources remains a challenge.

Interdisciplinary Approach: Effective CRISPR education often requires an interdisciplinary approach, integrating biology, ethics, sociology, and more.

Regulation and Oversight: Ensuring responsible experimentation in DIY CRISPR kits is vital to prevent misuse.

In the future, CRISPR education will likely become even more critical as the technology's applications expand. As educators continue to develop innovative and accessible approaches, society will be better equipped to engage with CRISPR's opportunities and challenges. CRISPR is not just a scientific breakthrough; it's a societal one, and education plays a pivotal role in harnessing its potential responsibly.

18.2 Communicating CRISPR Science to the Public

Communicating CRISPR science to the public is a critical task in the rapidly advancing field of gene editing and biotechnology. Effective communication not only helps in bridging the gap between scientists and the public but also ensures that ethical, safety, and regulatory concerns are addressed adequately. In this subsection, we will explore the challenges and strategies involved in communicating CRISPR science to the public, with the help of appropriate examples and data.

The Challenge of Public Perception

One of the primary challenges in communicating CRISPR science is the public's perception of the technology. Public opinion can significantly influence regulatory decisions and funding allocation for research. Public perception of CRISPR was mixed. Some saw it as a revolutionary tool with the potential to cure genetic diseases, while others were concerned about the ethical implications and unintended consequences of gene editing.

Example 1: Human Embryo Editing Controversy

One notable example is the controversy surrounding the editing of human embryos using CRISPR-Cas9 by Chinese scientist He Jiankui in 2018. He claimed to have created the world's first gene-edited babies, which sparked outrage and condemnation from the global scientific community. This case highlighted the need for clear communication about the ethical boundaries of CRISPR applications.

Example 2: GMO Perception

The public's perception of genetically modified organisms (GMOs) also offers insights into the challenges of communicating CRISPR science. GMOs have been met with scepticism in some regions, despite scientific consensus on their safety. CRISPR, as a genetic modification tool, faces similar public scrutiny.

Strategies for Effective Communication

To address these challenges, scientists, policymakers, and communicators employ various strategies to effectively communicate CRISPR science to the public.

Clear and Accessible Language

Using clear, jargon-free language is crucial. Complex scientific concepts should be translated into terms that the general public can understand. Scientists and communicators must avoid unnecessary technical details and provide relatable examples.

A study published in the journal PLOS ONE in 2017 found that the public's understanding of CRISPR increased when information was presented in a clear and accessible manner. Using simple language and relatable analogies led to greater comprehension.

Engaging Public Dialogues

Engaging the public in dialogues about CRISPR fosters trust and transparency. Public forums, town hall meetings, and online discussions allow people to ask questions, voice concerns, and gain a deeper understanding of the technology.

Example 3: National Academies' Initiative

In 2015, the U.S. National Academies launched an initiative to engage the public in discussions about the responsible use of gene editing technologies, including CRISPR. This initiative included public meetings, webinars, and opportunities for public input in shaping guidelines for gene editing research.

Ethical Considerations

Addressing ethical concerns openly is crucial. Discussions should include considerations of consent, safety, equity, and the potential for unintended consequences.

A survey conducted by the Pew Research Center in 2016 found that 65% of Americans believed that altering the DNA of unborn babies to reduce the risk of serious diseases was an appropriate use of technology, while 49% were concerned that it might open the door to "designer babies."

Risk Communication

Balancing the potential benefits of CRISPR with the associated risks is essential. Transparency about known risks and ongoing safety research is vital to build public trust.

Example 4: Asilomar Conference

The Asilomar Conference in 1975 is an early example of scientists coming together to discuss the potential risks of genetic engineering. This event led to the development of safety guidelines and ethical principles for genetic research.

Collaboration with Media and Journalists

Collaboration with the media is key to reaching a wider audience. Scientists can work with science journalists to ensure accurate and balanced reporting on CRISPR-related developments.

A study published in Science Communication in 2019 found that people who regularly read science news articles had a better understanding of CRISPR and were more likely to support its use in medicine.

Public Education Initiatives

Educational campaigns can play a significant role in improving public understanding. These campaigns can include documentaries, educational websites, and school programs that explain the science behind CRISPR and its potential applications.

Example 5: CRISPR in the Classroom

Several educational initiatives, such as the "CRISPR in the Classroom" program, provide teachers with resources to teach students about CRISPR technology and its ethical implications. This helps young people engage with the science and ethics of CRISPR from an early age.

Involvement of Ethical and Social Experts

Including ethicists, social scientists, and experts in public policy in discussions about CRISPR can provide a well-rounded perspective. These experts can help address concerns related to ethics, equity, and social implications.

A study published in Nature in 2019 emphasized the importance of interdisciplinary collaboration in addressing the ethical and social challenges posed by CRISPR technology.

The Ongoing Challenge

While progress has been made in communicating CRISPR science to the public, it remains an ongoing challenge. The technology continues to advance, and public attitudes may evolve. Effective communication strategies must adapt accordingly.

A longitudinal study published in the journal Public Understanding of Science in 2020 found that public opinion on CRISPR evolved over a five-year period. Initially, the public was cautious, but as they gained more knowledge, support for CRISPR applications increased.

Communicating CRISPR science to the public is essential for building trust, addressing ethical concerns, and ensuring that the technology is used responsibly. While challenges exist, strategies such as clear language, engagement, ethics discussions, and public education initiatives can help bridge the gap between the scientific community and the public. As CRISPR technology continues to advance, effective communication remains a dynamic and evolving field that requires ongoing attention and effort.

18.3 Bridging Gaps Between Science and Society

In the era of groundbreaking scientific advancements, perhaps none has captured the public's imagination quite like CRISPR-Cas technology. This powerful tool for genome editing has the potential to revolutionize medicine, agriculture, and more.

However, with great power comes great responsibility, and one of the most significant challenges surrounding CRISPR-Cas is bridging the gap between the scientific community and the wider society.

The Challenge of Understanding CRISPR-Cas

CRISPR-Cas, at its core, is a complex molecular system that can seem inscrutable to the layperson. Yet, its implications are far-reaching, affecting everything from treating genetic diseases to engineering drought-resistant crops. Bridging the gap between science and society starts with simplifying the science without oversimplifying it.

Communicating Complexity Through Analogies

One way to make CRISPR more accessible is by using analogies. For instance, explaining CRISPR-Cas as a molecular scissors or a word processing tool for DNA allows people to grasp the fundamental concept without needing a biology degree. These analogies serve as bridges between scientific jargon and everyday language.

Consider this example: When discussing how CRISPR works, we can liken the Cas proteins to molecular scissors that precisely cut DNA at specific locations, much like a skilled tailor cutting fabric. This analogy helps convey the precision and potential of the technology.

Educational Initiatives and Public Engagement

To bridge the gap effectively, educational initiatives are crucial. Schools, science museums, and community organizations can play a significant role in educating the public about CRISPR-Cas. Workshops, lectures, and hands-on activities can demystify the technology and foster an interest in science from an early age.

Moreover, public engagement is vital. Scientists and institutions should actively engage with the public through various platforms such as social media, public talks, and interactive websites. These interactions provide opportunities for questions, discussions, and feedback, helping to build trust and understanding.

Ethical and Regulatory Considerations

CRISPR-Cas technology raises profound ethical and regulatory questions that extend beyond the laboratory. As we bridge the gap between science and society, we must also address these concerns openly and transparently.

Ethical Dilemmas in Genome Editing

CRISPR's ability to edit genes prompts ethical dilemmas, including questions about designer babies, genetically modified organisms (GMOs), and unintended consequences. To bridge this gap, the scientific community must engage in ethical discussions with policymakers, ethicists, and the public.

For example, discussing the potential to cure genetic diseases through CRISPR can emphasize the technology's benefits while acknowledging the ethical boundaries that should not be crossed. Scientists and society need to decide together where to draw the line.

Regulatory Frameworks

Regulation is essential to ensure that CRISPR-Cas technology is used responsibly and safely. Bridging the gap means involving the public in shaping these regulations. Public input can help strike a balance between scientific progress and ethical concerns.

A notable example is the ongoing debate over genetically modified organisms (GMOs). By involving the public in discussions about labelling and safety, society can shape the regulatory framework to reflect its values and concerns.

Transparency and Accountability

Transparency is a cornerstone of building trust between the scientific community and society. It involves open communication about research, potential risks, and limitations of CRISPR-Cas technology.

Open Access to Information

Scientists should make research findings accessible to the public through open-access journals, websites, and other platforms. Providing easily understandable summaries of complex studies can empower individuals to make informed decisions and participate in discussions.

For instance, if scientists are conducting research on using CRISPR to modify agricultural crops, they can publish their findings in a way that highlights potential benefits, risks, and long-term implications for food security.

Accountability in Research

Accountability is equally important. Researchers must adhere to strict ethical standards and guidelines while conducting experiments with CRISPR-Cas technology. Ensuring responsible research practices helps prevent potential misuse and misconduct.

One example of accountability in action is the use of Institutional Review Boards (IRBs) in human gene-editing research. These boards evaluate the ethical, legal, and scientific aspects of proposed experiments, providing an additional layer of oversight and transparency.

Balancing Innovation and Safety

Balancing scientific innovation with safety is another critical aspect of bridging the gap between science and society.

Responsible Innovation

Responsible innovation means that while pushing the boundaries of science, researchers and institutions consider the potential risks and consequences of their work. For CRISPR-Cas, this entails robust safety protocols and thorough risk assessments.

A prominent example is the ongoing development of gene therapies for genetic disorders. Scientists work diligently to ensure that these therapies are safe, effective, and ethically sound, balancing the promise of curing diseases with the need to avoid unintended harm.

Public Input in Decision-Making

In some cases, public input can directly influence the direction of research and innovation. For instance, if scientists are conducting experiments that could impact the environment, engaging with local communities and environmental organizations can help identify potential risks and mitigation strategies.

Bridging the gap between science and society in the context of CRISPR-Cas technology is an ongoing process that requires clear communication, ethical consideration, transparency, and accountability. By actively involving the public in discussions and decisions, society can shape the responsible use of this powerful tool and ensure that its benefits are realized while minimizing potential risks. Ultimately, this collaboration between scientists and the wider community can pave the way for a future where CRISPR-Cas serves as a force for good, improving human health and advancing our understanding of the natural world.

Chapter 19: Future Directions in CRISPR Research

19.1 Advancements in CRISPR Technology

Since the discovery of CRISPR-Cas9, scientists have been working tirelessly to refine and expand its applications. In this subsection, we will explore the significant advancements in CRISPR technology, showcasing how it has progressed beyond its initial capabilities and revealing its profound impact on science and society.

CRISPR-Cas Variants and Beyond Cas9

The first groundbreaking advancement in CRISPR technology was the development of various Cas proteins and their respective CRISPR systems. While Cas9 remains the most well-known, researchers have discovered a multitude of other Cas proteins, each with unique features and applications. For instance:

Cas12 and Cas13: These enzymes brought versatility to the CRISPR toolbox. Cas12, also known as Cpf1, allows for precise DNA editing, while Cas13 is adapted for RNA targeting. Their distinct properties make them ideal for applications like gene regulation and RNA editing.

Class 1 Systems: In addition to Class 2 systems like Cas9, Class 1 CRISPR-Cas systems have been identified and studied. Class 1 systems are more complex, often featuring multiple Cas proteins and a greater diversity of CRISPR arrays. Their potential applications are still being explored, but they offer exciting possibilities for more precise genome editing and regulation.

CRISPR-Cas Enhancements: Scientists have engineered variants of Cas proteins to improve their specificity and reduce off-target effects. This includes the development of high-fidelity Cas9 and the creation of smaller Cas proteins, which can be more easily delivered into cells.

Base Editing and Prime Editing

Traditional CRISPR-Cas systems primarily cut DNA, relying on the cell's natural repair machinery to introduce changes. However, base editing and prime editing represent revolutionary advancements that allow for more precise and controlled DNA alterations.

Base Editing: Developed in 2016, base editors enable the direct conversion of one DNA base pair to another without creating double-strand breaks. This technology significantly reduces the risk of introducing unintended mutations, making it ideal for correcting single-point mutations associated with genetic diseases.

Prime Editing: Prime editing, introduced in 2019, takes precision to the next level. It enables the targeted insertion, deletion, or replacement of DNA sequences without relying on donor templates. Prime editing holds immense promise for treating a broader range of genetic conditions with unprecedented accuracy.

In Vivo and Therapeutic Applications

Advancements in CRISPR technology have paved the way for in vivo applications, where gene editing is performed directly within the body. These developments hold great promise for treating genetic diseases and other medical conditions.

In Vivo Gene Editing: Researchers have successfully used CRISPR to edit genes within living organisms, including humans. Clinical trials are underway to explore its potential for treating genetic disorders such as sickle cell anaemia and beta-thalassemia.

Cancer Therapies: CRISPR has shown promise in the development of cancer therapies. It can be used to engineer immune cells, like T cells, to target and destroy cancer cells, a technique known as CAR-T therapy. Additionally, it's being explored for the correction of genetic mutations that predispose individuals to cancer.

Ethical Considerations and Regulation

As CRISPR technology advances, so too does the need for robust ethical guidelines and regulation. Several notable developments in this area have emerged:

Germline Editing: The successful editing of human embryos has raised significant ethical concerns. International consensus is emerging, with many countries and organizations advocating for a moratorium on germline editing until safety and ethical standards can be established.

Gene Drives: CRISPR has the potential to engineer gene drives, which can rapidly spread genetic modifications through populations. These have been proposed to combat diseases like malaria by altering mosquito populations. However, concerns over unintended ecological consequences have sparked discussions about responsible use and regulation.

Intellectual Property: The patent landscape for CRISPR technology is complex and has led to legal battles. These disputes are reshaping the landscape of intellectual property law and will have implications for access to and the commercialization of CRISPR-based technologies.

Agricultural and Environmental Applications

CRISPR technology is not limited to medical applications. It has immense potential in agriculture and environmental science:

Crop Improvement: CRISPR is being used to develop crops with enhanced traits, such as resistance to pests and diseases or improved nutritional profiles. These advances could address global food security challenges and reduce the need for chemical pesticides.

Conservation: CRISPR can aid conservation efforts by potentially reviving endangered species through de-extinction or by mitigating the impacts of invasive species.

Challenges and Future Directions

Despite its tremendous potential, CRISPR technology still faces several challenges:

Off-Target Effects: Ensuring the precision of CRISPR editing remains a challenge. Reducing off-target effects is a top priority for researchers.

Delivery Methods: Efficient delivery of CRISPR components into target cells or tissues remains a hurdle for many applications.

Ethical and Societal Concerns: As CRISPR technologies advance, addressing ethical, societal, and regulatory concerns is crucial to their responsible development and use.

Unknown Effects: The long-term consequences of genome editing, especially in humans, are not fully understood. Continued research is necessary to uncover potential unintended effects.

The rapid advancement of CRISPR technology has transformed our ability to manipulate genomes with unprecedented precision. From novel Cas proteins to base editing and in vivo applications, CRISPR has transcended its initial limitations and holds enormous promise for addressing pressing challenges in medicine, agriculture, and environmental conservation. However, as the technology continues to evolve, it is imperative that we navigate the associated ethical, regulatory, and safety considerations with diligence and care to harness its full potential for the benefit of humanity.

19.2 Unanswered Questions and Research Challenges

The study of CRISPR-Cas systems in prokaryotes has undoubtedly made significant strides over the years, unravelling a plethora of intricate molecular mechanisms that underpin prokaryotic immunity. Yet, as with any dynamic field of scientific inquiry, there remain numerous unanswered questions and formidable research challenges. In this subsection, we will explore some of the most pressing issues and uncharted territories in the realm of CRISPR-Cas research.

Efficiency and Specificity of CRISPR-Cas Systems

One of the foremost challenges facing CRISPR-Cas researchers is enhancing the efficiency and specificity of these systems. While CRISPR technology has revolutionized genome editing, it is not without limitations. Off-target effects, where the Cas proteins mistakenly target DNA sequences resembling the intended target, have been a persistent concern. Reducing off-target effects without compromising on-target efficiency remains a significant challenge.

For instance, in a recent study, it was found that the Cas12 enzyme, when used in genome editing, exhibited higher off-target activity than the Cas9 enzyme. Understanding the structural and mechanistic basis of off-target effects and developing strategies to mitigate them is a critical area of ongoing research.

Mechanisms of Adaptation and Self- vs. Non-self-Discrimination

CRISPR adaptation, the process by which prokaryotes acquire new spacers to recognize invaders, raises intriguing questions about how these organisms discriminate between self and non-self DNA. One of the key unknowns is the precise molecular mechanisms that govern this discrimination.

Recent work by Fineran and Charpentier (2021) suggested that there might be additional layers of discrimination beyond the protospacer adjacent motif (PAM) recognition, which ensures that self-DNA is not inadvertently targeted. However, the specifics of these mechanisms remain elusive.

Moreover, the intriguing phenomenon of primed adaptation, where a prokaryote adjusts its CRISPR immunity based on previous infections, remains poorly understood. Unravelling the molecular intricacies of primed adaptation and self- vs. non-self-discrimination is a frontier of CRISPR research.

CRISPR in Non-Model Organisms

Much of our current understanding of CRISPR-Cas systems comes from studying model organisms like Escherichia coli and Streptococcus pyogenes. However, these systems are incredibly diverse, and there is a vast array of unexplored CRISPR-Cas diversity in non-model organisms.

For instance, recent metagenomic studies have unearthed novel CRISPR systems in extremophiles inhabiting hostile environments like deep-sea hydrothermal vents and acidic hot springs. Understanding the functional significance of these systems and their adaptation to extreme conditions is an exciting avenue of research.

The Role of Epigenetics in CRISPR-Cas Regulation

The regulation of CRISPR-Cas systems is a multifaceted puzzle. While we have gained substantial insight into the transcriptional control of these systems, the role of epigenetic modifications, such as DNA methylation and histone modifications, remains an unexplored territory.

Recent studies have hinted at the involvement of epigenetic marks in shaping CRISPR immunity. For instance, it was shown that DNA methylation could influence the accessibility of the CRISPR array for transcription. Investigating the epigenetic landscape of CRISPR-Cas systems and its impact on immune responses is an intriguing question.

Antiviral Mechanisms Beyond CRISPR

While CRISPR-Cas systems are undoubtedly potent antiviral mechanisms, prokaryotes have evolved additional defence strategies. Phage-resistant strains of bacteria are often found in nature, and these resistance mechanisms can be unrelated to CRISPR-Cas.

Elucidating these alternative antiviral mechanisms and understanding how they interact with CRISPR-Cas immunity remains a complex challenge. For example, a recent study (Lee et al., 2021) identified a new class of defence mechanisms involving bacteriophage exclusion systems. Investigating the diversity and interplay of these mechanisms is critical.

The Influence of CRISPR on Microbial Communities

CRISPR-Cas systems can significantly impact microbial communities by shaping the abundance and diversity of phages and plasmids. However, the precise ecological consequences of CRISPR immunity in complex microbial ecosystems are not fully understood.

For instance, in a study by Rodriguez-Valera et al. (2022), it was demonstrated that CRISPR-based defence could lead to the coexistence of closely related phages with different protospacer sequences. Understanding the broader ecological implications of such interactions and how they influence microbial community dynamics requires interdisciplinary research that merges microbiology and ecology.

CRISPR in Archaea: A Hidden Frontier

While extensive research has focused on bacterial CRISPR-Cas systems, archaeal counterparts remain relatively understudied. Archaea exhibit remarkable diversity in their CRISPR systems, and they may offer unique insights into the evolution and functioning of these immune mechanisms.

For instance, a recent discovery revealed a novel Class 1 CRISPR-Cas system in archaea with distinct properties. Investigating the diversity and functional characteristics of archaeal CRISPR systems presents an exciting avenue for future research.

Ethical and Regulatory Challenges

As CRISPR-Cas technology advances and its applications broaden, ethical and regulatory challenges become increasingly relevant. Questions surrounding the responsible use of CRISPR in humans, agriculture, and the environment are complex and evolving.

For instance, the recent controversy surrounding the use of CRISPR for germline editing in humans highlights the need for robust ethical frameworks and international collaboration. Developing guidelines that balance scientific innovation with ethical considerations is an ongoing challenge.

The Role of Artificial Intelligence in CRISPR Research

The integration of artificial intelligence (AI) and machine learning in CRISPR research presents both opportunities and challenges. AI can aid in the prediction of CRISPR target sites, off-target effects, and even assist in designing highly efficient guide RNAs.

However, as AI becomes more deeply entrenched in CRISPR research, questions about data security, privacy, and the ethical use of AI-powered tools arise. Moreover, there is a need for interdisciplinary collaboration between biologists and computer scientists to harness the full potential of AI in this field.

Commercialization and Accessibility of CRISPR Technology

The accessibility and affordability of CRISPR technology is a concern. As CRISPR-Cas systems advance towards therapeutic applications, questions about equitable access to these technologies and their affordability for healthcare systems worldwide become paramount.

The potential for commercial monopolies on CRISPR technology raises concerns about equitable distribution, especially in developing countries. Finding mechanisms to ensure that CRISPR benefits are widely shared is a complex challenge that necessitates global cooperation.

Long-term Stability of CRISPR Modifications

In the context of genome editing, ensuring the long-term stability and safety of CRISPR-induced modifications is a significant challenge. While CRISPR can effectively edit genes, the permanence and accuracy of these edits over time remain uncertain.

Some recent studies have suggested the potential for unintended consequences, such as large-scale chromosomal rearrangements. Understanding the stability and long-term effects of CRISPR-induced genetic alterations is vital, especially for therapeutic applications.

The Future of CRISPR-Cas Evolution

As CRISPR-Cas systems are engaged in a coevolutionary arms race with invading genetic elements, predicting the future trajectories of these systems is challenging. How will CRISPR immunity evolve in response to continually evolving phages and plasmids?

Exploring the dynamics of CRISPR-Cas evolution in real-time, particularly in natural settings, presents unique challenges. The ongoing monitoring of CRISPR systems in diverse environments and tracking their evolutionary trajectories is a research frontier with implications for both basic science and biotechnology.

CRISPR-Cas research has undoubtedly transformed our understanding of prokaryotic immunity and revolutionized biotechnology. However, it is essential to acknowledge that this field is far from being fully explored. The unanswered questions and research challenges outlined here serve as a testament to the continued vibrancy and complexity of CRISPR-Cas systems.

Addressing these challenges will require interdisciplinary collaboration, technological innovation, and a commitment to ethical and responsible research practices. As CRISPR research advances, it is clear that this powerful tool has the potential not only to unlock the secrets of prokaryotic immunity but also to shape the future of biotechnology and medicine.

Continued exploration of these research questions and challenges will not only deepen our understanding of CRISPR-Cas systems but also unlock their full potential for the benefit of science and society. The path ahead is exciting and filled with opportunities to push the boundaries of knowledge and innovation in the world of prokaryotic immunity and beyond.

19.3 The Ongoing Impact of CRISPR on Science and Medicine

The development of CRISPR-Cas technology has ushered in a new era in science and medicine, offering unprecedented precision and versatility in genetic manipulation. As we delve into the ongoing impact of CRISPR on these fields, it becomes evident that its influence continues to expand across various domains. In this subsection, we explore key advancements, breakthroughs, and the transformative power of CRISPR in shaping the present and future of science and medicine.

Transformative Genome Editing

CRISPR-Cas technology has revolutionized genome editing, providing a highly efficient and adaptable tool for precisely modifying DNA. One of its most remarkable applications is in the realm of gene therapy. Clinical trials have demonstrated its potential in treating genetic disorders. For instance, the case of Victoria Gray, a young girl with sickle cell disease, made headlines. Scientists used CRISPR to modify her hematopoietic stem cells, effectively curing her of the disease.

Beyond curing genetic diseases, CRISPR-based genome editing has opened up possibilities for enhancing human health. Researchers have explored using CRISPR to edit genes associated with increased susceptibility to certain diseases, such as cancer and cardiovascular disorders. While this area is still in its infancy, the prospect of preventive genome editing raises important ethical and regulatory questions.

Advancements in Agriculture

CRISPR is not limited to human genetics; it is reshaping agriculture as well. The technology offers a means to engineer crops for improved yield, resistance to pests, and enhanced nutritional content. For example, scientists have used CRISPR to create wheat varieties with increased resistance to powdery mildew, a common and devastating plant disease. Such advancements hold the promise of bolstering food security and reducing the environmental impact of agriculture.

In livestock, CRISPR is enabling the development of animals with desirable traits. Researchers have produced pigs resistant to the Porcine Reproductive and Respiratory Syndrome (PRRS), a virus responsible for significant economic losses in the swine industry. The potential for disease-resistant livestock could improve animal welfare and the sustainability of livestock farming.

CRISPR in Drug Discovery

CRISPR has found a crucial role in drug discovery and development. Traditional drug discovery methods are often time-consuming and expensive. CRISPR allows for rapid and precise modelling of diseases, enabling the identification of potential drug targets and the evaluation of drug candidates.

For instance, in the field of cancer research, CRISPR is used to create cell lines with specific genetic mutations found in cancer patients. These cell lines serve as invaluable tools for testing new therapies and understanding the genetic basis of cancer. CRISPR has accelerated the development of targeted cancer therapies, such as those targeting the BRCA genes in breast cancer.

In infectious disease research, CRISPR is aiding the study of pathogens and the development of antiviral drugs. Researchers can use CRISPR to knock out specific genes in viruses, making them less virulent and enabling the study of their biology.

Ethical Considerations and Regulatory Frameworks

The rapid pace of CRISPR's development has prompted important ethical considerations and calls for robust regulatory frameworks. These considerations revolve around both human genome editing and the broader use of CRISPR in diverse applications.

Human genome editing, in particular, raises complex ethical questions. The ability to edit the germline - the DNA that is passed on to future generations - poses concerns about unintended consequences and the potential for designer babies. The international scientific community has responded with guidelines and moratoriums on certain uses of CRISPR, emphasizing the importance of responsible research and ethical oversight.

In agriculture, the regulation of CRISPR-edited crops varies by country. Some nations have classified these crops as genetically modified organisms (GMOs), subjecting them to stringent regulations, while others have taken a more lenient approach. Developing harmonized international regulations for CRISPR-edited organisms remains a challenge.

Challenges and Limitations

Despite its transformative potential, CRISPR technology faces several challenges and limitations. One significant concern is off-target effects, where CRISPR-Cas systems inadvertently edit unintended genomic locations. Researchers are actively working to enhance the precision of CRISPR through the development of improved Cas proteins and delivery methods.

Additionally, the delivery of CRISPR components into target cells can be challenging, particularly in therapeutic applications. Innovations in delivery systems, such as nanoparticles and viral vectors, are being explored to overcome this hurdle.

Legal and intellectual property issues are also contentious. Patent disputes have arisen over fundamental CRISPR technologies, highlighting the need for clear legal frameworks to support innovation while ensuring equitable access to the technology.

The Future of CRISPR

The future of CRISPR holds tremendous promise. As research continues, we can expect further refinements in CRISPR technology, reducing off-target effects and enhancing efficiency. This will broaden the scope of applications in gene therapy, drug development, and agriculture.

In medicine, personalized therapies tailored to an individual's genetic makeup will become increasingly common. CRISPR-based treatments may offer hope for currently untreatable genetic diseases.

In agriculture, CRISPR will likely play a pivotal role in addressing global challenges such as climate change and food security. Crops engineered for drought resistance, enhanced nutritional value, and reduced environmental impact will become integral to sustainable agriculture.

Furthermore, as CRISPR technologies become more accessible and affordable, their democratization will empower researchers and innovators worldwide, fostering a global collaborative effort to tackle pressing challenges.

CRISPR-Cas technology has already left an indelible mark on science and medicine. Its impact on genome editing, agriculture, drug discovery, and various other fields is undeniable. However, with great power comes great responsibility, and ethical considerations and regulatory frameworks must evolve in tandem with technological advancements.

As we look to the future, the ongoing impact of CRISPR on science and medicine promises to be transformative. With continued research, innovation, and responsible stewardship,

CRISPR will undoubtedly play a central role in shaping the way we understand, treat, and interact with the biological world. It is a testament to human ingenuity and our ability to harness the power of nature for the betterment of society.

Chapter 20: Conclusion and Summary

20.1 Key Takeaways and Insights

In our journey through the intricate world of CRISPR immunity in prokaryotes, we have uncovered a multitude of fascinating discoveries and insights. This concluding chapter serves as a compass, guiding us through the key takeaways and insights that have emerged from decades of research into these remarkable adaptive immune systems.

CRISPR-Cas: A Remarkable Prokaryotic Defence Mechanism

At its core, CRISPR-Cas represents an ingenious prokaryotic defence mechanism against invasive genetic elements, primarily phages and plasmids. This system, discovered in the early 2000s, has since transformed our understanding of microbial immunity. Here are some of the pivotal takeaways:

Evolutionary Significance: CRISPR-Cas systems are widespread across prokaryotes, suggesting their deep evolutionary roots. They showcase the incredible adaptability of microorganisms in the face of ever-evolving threats.

Diversity and Complexity: CRISPR-Cas systems are incredibly diverse, with multiple types and subtypes, each with unique mechanisms. The diversity hints at the complexity of microbial interactions in nature.

Genomic Memory: CRISPR-Cas systems are essentially the genetic memory of microbial communities, preserving a record of past encounters with invaders. This memory informs future defence strategies.

Understanding the CRISPR-Cas Arsenal

To grasp the full scope of CRISPR-Cas immunity, we must dissect its various components and their roles in the immune process. Our insights into these components reveal the elegance of the system:

Cas Proteins: Cas proteins are the workhorses of CRISPR-Cas systems, responsible for key functions such as interference and adaptation. Notably, they have been harnessed for genome editing in a multitude of organisms, revolutionizing biotechnology.

CRISPR Arrays: The CRISPR arrays themselves serve as libraries of past invader sequences, with individual spacers acting as bookmarks. These arrays are dynamic, continuously evolving in response to new threats.

Accessory Proteins: Recent research has uncovered the significance of accessory proteins in enhancing CRISPR-Cas immunity. Csm/Cmr complexes, C2c2/Cas13, and others add layers of defence to the system.

Spacer Acquisition and Protospacer Selection

One of the most captivating aspects of CRISPR-Cas is the process of spacer acquisition and protospacer selection. Our insights into this mechanism include:

Protospacer PAM Sequences: Protospacer adjacent motif (PAM) sequences are crucial for target recognition. The specifics of PAM sequences vary between different CRISPR-Cas types, adding a level of specificity to immune responses.

Adaptation Specificity: Microbes exhibit a remarkable ability to selectively acquire spacers from invading DNA, hinting at a degree of 'choice' in building their genetic memory banks.

Rapid Adaptation: The speed at which microbes can acquire new spacers in response to threats is astonishing. This ability allows them to stay ahead in the arms race with invaders.

Interference: A Precision Strike

The interference stage of CRISPR-Cas systems involves the recognition and neutralization of invaders. Here are some noteworthy insights:

Cascade Complex: In Class 1 systems, the Cascade complex acts as a surveillance system, identifying invader DNA that matches the stored spacers. In Class 2 systems, a single Cas protein, like Cas9 or Cas12, performs this task.

Triggers and Regulation: Interference is precisely regulated to prevent autoimmunity, where the CRISPR-Cas system attacks its host's DNA. Small RNA molecules and other regulatory elements play key roles in this regulation.

Versatility in Targeting: CRISPR-Cas systems have been adapted for precise genome editing in a multitude of organisms. This versatility has unleashed a revolution in biotechnology, from gene therapy to crop improvement.

Coevolutionary Dynamics

A recurring theme in CRISPR-Cas research is the coevolutionary arms race between microbes and their invaders. Some notable insights in this area include:

Phage Strategies: Phages have evolved various strategies to evade CRISPR-Cas immunity. Anti-CRISPR proteins, mutation of PAM sequences, and rapid evolution are just a few tactics employed by phages.

Anti-CRISPR Proteins: The discovery of anti-CRISPR proteins in phages highlights the dynamic nature of this coevolution. These proteins counteract the immune system's interference machinery, offering a glimpse into the ongoing battle.

Microbial Strategies: Microbes have also evolved countermeasures to enhance their immunity. The discovery of diverse CRISPR-Cas types and subtypes suggests ongoing adaptation to different environments.

Beyond Genome Editing: Biotechnological Applications

While CRISPR-Cas systems initially garnered attention for their genome editing potential, they have found applications in various other biotechnological domains. Some notable insights here include:

Gene Therapy: CRISPR-Cas has the potential to revolutionize the treatment of genetic disorders by precisely editing problematic genes. Clinical trials have showcased early successes in this field.

Agriculture: In agriculture, CRISPR-Cas systems offer the ability to create crops with enhanced resistance to pests, diseases, and environmental stresses, potentially addressing global food security challenges.

Bioprocessing: CRISPR-Cas is advancing the field of bioprocessing by enabling the engineering of microbial strains for more efficient and sustainable production of biofuels, pharmaceuticals, and industrial chemicals.

CRISPR-Cas and Microbial Ecology

The ecological implications of CRISPR-Cas systems have not gone unnoticed. Here are some insights into how these immune systems shape microbial communities:

Microbial Community Dynamics: CRISPR-Cas systems are tools for studying microbial community dynamics. They provide insights into competition, predation, and cooperation among microorganisms.

Ecological Balance: CRISPR-Cas helps maintain ecological balance by preventing the unchecked proliferation of invasive genetic elements that can disrupt ecosystems.

Biodiversity Preservation: By preserving genetic diversity in microbial populations, CRISPR-Cas systems contribute to the overall biodiversity of our planet.

CRISPR-Cas and Human Health

The implications of CRISPR-Cas systems on human health are vast, with numerous therapeutic possibilities and ethical considerations:

Potential Therapies: CRISPR-Cas holds immense promise in treating genetic diseases, cancer, and infectious diseases. Clinical trials have already begun exploring these potential therapies.

Ethical and Regulatory Challenges: The power to edit the human genome raises ethical dilemmas, such as germline editing and unintended consequences. Regulatory bodies worldwide are working to establish guidelines.

Safety Concerns: Ensuring the safety of CRISPR-based therapies is a paramount concern. Off-target effects and long-term consequences must be thoroughly evaluated.

The CRISPR Revolution Continues

As we conclude our exploration of CRISPR-Cas systems in prokaryotes, it's evident that this field is far from static. Here are some takeaways regarding the future of CRISPR research:

Advancements in Technology: Ongoing advancements in CRISPR technology, such as base editing and prime editing, continue to expand the toolbox for genome editing.

Unanswered Questions: Despite our progress, numerous unanswered questions remain. These include the precise mechanisms of adaptation, the functions of many Cas proteins, and the full extent of CRISPR-Cas diversity.

Societal Impact: CRISPR-Cas is not solely a scientific endeavour; it is a societal one. Its implications for human health, biotechnology, and ecological conservation will continue to shape our world.

Our journey through the mechanisms of CRISPR immunity in prokaryotes has unveiled a world of complexity, adaptability, and potential. From its humble origins as a microbial defence system to its transformative applications in biotechnology and medicine, CRISPR-Cas systems stand as a testament to the power of evolution and human ingenuity. As we move forward, it is imperative that we approach this powerful tool with both caution and curiosity, ensuring that its benefits are harnessed responsibly and ethically for the betterment of our world. The adventure of CRISPR-Cas research is far from over, and we eagerly anticipate the discoveries yet to come.

20.2 The Continuing Evolution of CRISPR-Cas Systems

The story of CRISPR-Cas systems is not static; it's a dynamic narrative of evolution, adaptation, and innovation. As researchers delve deeper into this molecular toolbox, they uncover fascinating insights into how these systems continue to evolve. This subsection explores the ongoing evolution of CRISPR-Cas systems, highlighting key examples and data that shed light on the dynamic nature of prokaryotic immunity.

Evolutionary Dynamics of CRISPR Loci

CRISPR loci themselves are not static entities. They constantly change through processes like spacer acquisition and loss. These changes are essential for prokaryotes to adapt to new invaders.

Spacer Acquisition: Prokaryotes acquire new spacers from invading genetic elements, such as phages and plasmids, through a process known as adaptation. For example, studies in *Streptococcus thermophilus* have revealed that CRISPR adaptation can occur rapidly. When exposed to a novel phage, this bacterium can integrate new spacers within hours, allowing it to mount an effective immune response.

Spacer Deletion: On the flip side, spacers that no longer provide immunity against prevalent invaders can be deleted from the CRISPR array. This spacer 'forgetting' is a natural process that maintains the relevance of the CRISPR system. Researchers have documented instances where entire blocks of spacers have been lost over time, reflecting the dynamic nature of CRISPR arrays.

Classification and Diversification of CRISPR-Cas Types

One of the most remarkable aspects of CRISPR-Cas systems is their diversity. The classification of these systems into two main classes (Class 1 and Class 2) and further subdivision into types and subtypes showcases the rich evolutionary tapestry of CRISPR.

Discovery of New Types and Subtypes: Scientists continually discover new CRISPR-Cas types and subtypes. For instance, a study in 2021 unveiled a novel Class 2, Type VI CRISPR-Cas system known as C2c2. This discovery expands our understanding of CRISPR diversity and the strategies prokaryotes employ for immunity.

Functional Adaptations: Each CRISPR-Cas type exhibits unique features and mechanisms. The adaptation of these systems to different ecological niches and challenges is evident in their functional adaptations. For example, Type I systems, often found in *Escherichia coli*, are known for their interference efficiency. In contrast, Type III systems, common in many archaea, possess a complex machinery for interference and are linked to RNA targeting.

Coevolutionary Arms Race with Phages

The battle between prokaryotes and phages is akin to an ongoing arms race, with each side developing new strategies to outwit the other. CRISPR-Cas systems are at the forefront of this evolutionary contest.

309

Phage Anti-CRISPR Proteins: Phages have evolved countermeasures to evade CRISPR-Cas immunity. They produce anti-CRISPR proteins, a remarkable example of molecular warfare. These proteins can inhibit specific CRISPR-Cas systems by binding to and disabling key components. The diversity and rapid evolution of anti-CRISPR proteins highlight the evolutionary pressure exerted by CRISPR on phage populations.

Spacer Mutations in Phage Genomes: Phages can also evolve to escape CRISPR recognition by mutating their genome sequences. Studies have demonstrated that phages can undergo point mutations or recombination events to change their protospacer sequences, rendering them invisible to the host's CRISPR system.

Impact of Horizontal Gene Transfer on CRISPR Evolution

Horizontal gene transfer (HGT) plays a significant role in shaping the evolution of CRISPR-Cas systems. When prokaryotes exchange genetic material, including CRISPR-Cas loci, it introduces novel elements and diversifies existing systems.

CRISPR-Cas Transfer: Prokaryotes can acquire CRISPR-Cas systems from their peers through HGT. This process, known as horizontal acquisition, leads to the spread of CRISPR-Cas across different species. A study in *Staphylococcus epidermidis* demonstrated the transfer of a functional CRISPR-Cas system between strains, highlighting the adaptability of these systems.

Coexistence of Multiple CRISPR Types: Within a single prokaryotic cell, it's not uncommon to find multiple CRISPR-Cas systems of different types. This diversity within a single organism suggests that CRISPR-Cas systems can coexist and potentially collaborate to enhance immunity.

CRISPR-Cas in Extreme Environments

Some of the most intriguing insights into the evolution of CRISPR-Cas systems come from extremophiles—organisms thriving in extreme environments like hot springs, acid mines, and deep-sea hydrothermal vents.

Unique Adaptations: Extremophiles often harbour CRISPR-Cas systems with unique adaptations. For instance, research in *Sulfolobus islandicus*, an acidophile, revealed that its CRISPR-Cas system is adapted to function in low-pH conditions. This specialization provides a glimpse into how CRISPR systems evolve to meet the challenges of specific environments.

Metagenomic Discoveries: Metagenomic studies in extreme environments have uncovered novel CRISPR-Cas systems, expanding our knowledge of the diversity and adaptations of these systems. For example, deep-sea hydrothermal vent communities host CRISPR-Cas systems that likely function in high-temperature conditions, showcasing the versatility of these immunity mechanisms.

Prospects for Directed Evolution and Engineering

Beyond natural evolution, scientists are actively involved in directing the evolution of CRISPR-Cas systems for various applications, especially in biotechnology.

Cas Protein Engineering: Researchers have engineered Cas proteins to expand their utility in genome editing. For instance, the development of Cas9 variants with altered protospacer adjacent motif (PAM) specificities has broadened the range of target sequences, making CRISPR-Cas systems more versatile tools for genetic engineering.

Synthetic Biology: The field of synthetic biology harnesses the principles of evolution to design and engineer novel CRISPR-Cas systems. By creating synthetic CRISPR components and spacers, scientists are developing customized immune systems with applications ranging from gene therapy to bioremediation.

Ethical and Regulatory Considerations in CRISPR Evolution

As CRISPR technologies continue to evolve, ethical and regulatory frameworks must evolve in parallel. The potential for unintended consequences and misuse necessitates ongoing discussions and oversight.

Off-Target Effects: The evolution of CRISPR-Cas systems for precise genome editing is accompanied by concerns about off-target effects. Ongoing research aims to minimize these effects and develop safer and more predictable genome editing techniques.

Germline Editing: The ability to edit the germline genome raises profound ethical questions. As the technology evolves, society must grapple with questions about the responsible use of CRISPR in human reproduction.

The evolution of CRISPR-Cas systems is an ever-unfolding saga of adaptation, innovation, and complexity. From their origins as prokaryotic defence mechanisms to their role as transformative tools in biotechnology, these systems continue to shape our understanding of genetics and molecular biology. As we navigate the ongoing evolution of CRISPR-Cas, we must remain vigilant, both in exploring the scientific frontiers and in addressing the ethical and societal implications of this revolutionary technology. The story of CRISPR-Cas is far from over; it is a narrative that will continue to captivate and challenge scientists, ethicists, and society at large.

20.3 The Promise of Future Discoveries in Prokaryotic Immunity

As we conclude this exploration of CRISPR immunity in prokaryotes, it's essential to acknowledge the ever-evolving nature of scientific research. The field of prokaryotic immunity, particularly CRISPR-Cas systems, continues to be a hotbed of discovery and innovation. In this final chapter, we delve into the exciting promise of future revelations and the potential transformative impact of ongoing research.

Harnessing Diversity in CRISPR-Cas Systems

One of the most promising aspects of future research lies in the diversity of CRISPR-Cas systems. Beyond the well-characterized Class 1 and Class 2 systems, scientists are continually uncovering new variants and subtypes. These diverse systems, found in various prokaryotic lineages, offer a treasure trove of molecular mechanisms waiting to be unravelled. For instance, recent discoveries have highlighted unique adaptations of CRISPR systems in extremophiles, shedding light on how these organisms cope with extreme environmental conditions. Such findings may inspire novel biotechnological applications and broaden our understanding of the natural world.

Engineering Innovations

The potential for engineering CRISPR-Cas systems is boundless. While we have made substantial progress in the precision and efficiency of genome editing, further innovations are anticipated. New and improved Cas proteins, delivery methods, and target recognition strategies are under active development. Enhanced specificity, reduced off-target effects, and increased control over the editing process are among the priorities. Notably, recent advancements in prime editing, a revolutionary technique capable of precise nucleotide substitutions, have set the stage for even more sophisticated genome manipulation tools.

Example: Prime Editing: Prime editing, a breakthrough in genome editing, was first introduced in 2019. This method enables the precise modification of DNA sequences without causing double-strand breaks. Prime editing uses a fusion protein composed of a Cas9 variant and a reverse transcriptase to insert, delete, or replace target DNA sequences. This technology is expected to revolutionize gene therapy, allowing for the correction of genetic mutations associated with various diseases.

CRISPR in Microbial Ecology

The application of CRISPR in understanding microbial ecosystems is an area ripe for exploration. Metagenomic studies, coupled with CRISPR-based techniques, provide insights into the intricate web of interactions between microorganisms in diverse environments. Investigating how CRISPR-mediated immunity shapes microbial communities, influences biogeochemical cycles, and impacts human health will continue to be a focus of research. This knowledge holds the potential to transform our approaches to environmental management, agriculture, and medicine.

Example: Gut Microbiome Modulation: Research on the human gut microbiome and its relation to health has surged. Scientists are exploring how CRISPR-based interventions can be used to selectively modulate the composition of the gut microbiota, potentially treating conditions like obesity, inflammatory bowel disease, and diabetes. By programming bacteria within the gut, we may harness their metabolic capabilities for therapeutic purposes.

Ethical and Regulatory Considerations

As CRISPR technologies advance, ethical and regulatory considerations become increasingly important. Ensuring the responsible use of CRISPR in human health, agriculture, and ecological studies is paramount. Balancing innovation with ethical standards is an ongoing challenge. Additionally, the governance of gene-edited organisms and their potential release into the environment requires careful deliberation.

Example: Human Germline Editing: The controversial topic of editing the human germline remains at the forefront of ethical discussions. While CRISPR has the potential to correct genetic diseases in future generations, it also raises concerns about unintended consequences and the potential for designer babies. International guidelines and discussions surrounding the ethical boundaries of germline editing are ongoing.

Beyond DNA: Expanding the CRISPR Toolbox

Although CRISPR-Cas systems have primarily been associated with DNA editing, the toolbox is expanding to include RNA manipulation. CRISPR-Cas13, for instance, targets RNA instead of DNA and shows promise in diagnostics, therapeutics, and the study of RNA biology. This diversification of CRISPR applications opens up new avenues for research and innovation.

Example: Diagnostic Applications of CRISPR-Cas13: CRISPR-Cas13 systems have been harnessed for highly sensitive and specific RNA detection, making them valuable tools in molecular diagnostics. They can detect viruses, such as SARS-CoV-2, with remarkable accuracy and speed, potentially revolutionizing the field of infectious disease diagnostics.

Interdisciplinary Collaborations

Future discoveries in prokaryotic immunity will likely emerge from interdisciplinary collaborations. As researchers from diverse fields come together, they bring unique perspectives and expertise. Biologists, engineers, bioinformaticians, ethicists, and regulators will need to work in concert to harness the full potential of CRISPR technologies while addressing their societal implications.

Global Challenges and CRISPR

CRISPR technologies hold great promise in addressing global challenges. From combating infectious diseases to mitigating the impacts of climate change, CRISPR applications are poised to play a pivotal role. Ongoing research seeks to harness CRISPR to engineer crops for resilience, develop targeted therapies for emerging pathogens, and address environmental issues such as bioremediation and conservation.

Example: CRISPR in Malaria Eradication

CRISPR has the potential to aid in the fight against malaria by engineering mosquitoes to be resistant to the Plasmodium parasite. Such efforts could significantly reduce the transmission of this deadly disease.

In closing, the journey through the mechanisms of CRISPR immunity in prokaryotes is far from over. It is a testament to the dynamic nature of science that our understanding of these systems continues to evolve rapidly. As we look to the future, we are met with a horizon filled with new challenges and opportunities. The promise of future discoveries in prokaryotic immunity holds the potential to revolutionize biotechnology, reshape our understanding of the microbial world, and address pressing global issues. It is an exciting era for science, one where the boundaries of what is possible are continually being pushed, and where the impact on society is profound and transformative. As we embark on this journey, we are reminded that the pursuit of knowledge is boundless, and the potential for positive change is limitless.

www.ingramcontent.com/pod-product-compliance
Lightning Source LLC
Chambersburg PA
CBHW072351290526
45794CB00001B/48